THE ATOMIC CHEF

And Other True Tales of Design, Technology, and Human Error

THE ATOMIC CHEF

And Other True Tales of Design, Technology, and Human Error

STEVEN CASEY

Aegean

Aegean Publishing Company
Santa Barbara

Published in the U.S.A. by
Aegean Publishing Company

Post Office Box 6790, Santa Barbara, California 93160

THE ATOMIC CHEF
AND OTHER TRUE TALES
OF DESIGN, TECHNOLOGY, AND HUMAN ERROR

ISBN 10: 0-9636178-6-9 (hc)
ISBN 13: 978-0-9636178-6-6 (hc)

If not available from your local bookstore, this book may be ordered directly from the publisher. Send the cover price plus $4.50 for shipping and handling to the above address. California residents add applicable sales tax.

Publishers Cataloging in Publication Data
Casey, Steven M.
The atomic chef
and other true tales of design, technology, and human error.

p. cm.
Includes bibliographical references
ISBN 978-0-9636178-6-6 (hc) : $29.00
1. Technology. 2. Human engineering.
3. Human factors 4. Engineering design. 5. Title
TA 166 C37 2006 620.8′2 2005-932971

TABLE OF CONTENTS

People are the quintessential element
in all technology...

Once we recognize
the inescapable human nexus of all technology
our attitude toward the reliability problem
is fundamentally changed.

Garrett Hardin
1915 - 2003

PROLOGUE

A true story told from the perspective of a person overtaken by surprising yet foreseeable events is a compelling way to learn about user interface design and the safety of technologies on which we have become so dependent. *The Atomic Chef* — like its forerunner book, *'Set Phasers on Stun' and Other True Tales of Design, Technology, and Human Error* — contains 20 new stories about people interacting unsuccessfully with technology in settings that are as varied as the environments and equipment designed and used by human beings. By focusing on the plight of the *user*, it is my hope to convey to designers, managers, system operators, and the general public how incompatibilities between the way things are designed and the way people perceive, think, and act can result in human error, or, more accurately, *design-induced error*.

The integrity of the interface between people and the objects they use is as important as the soundness of any physical part, electronic component, or line of software code. As technologies mature and become more reliable overall, the source of accidents often shifts from being hardware-related to interface-related. The discipline of *ergonomics*, or *human factors engineering*, seeks to address human characteristics, capabilities, and limitations and reflect them in the design of the things we create to make them easier to use, more reliable, and safer.

This collection of stories was selected because I found each intriguing, relevant, and, ultimately, instructive. Some of these tales are intended to be humorous — but every bit as educational as the stories that end tragically. I have also deliberately selected stories from a variety of settings, locations, and points in time. Design-induced human error is, after all, not restricted to any single technology or continent.

As in *Set Phasers on Stun*, the stories in this new book also represent different classes and types of human error. The chapter *Negative Transfer*, for example, is a classic tale of what ergonomists call, appropriately, "negative transfer," and illustrates the importance of user expectancies and consistency in control design in an experimental NASA aircraft. A vignette about roadway signage and its importance to the traveler is told in the true tale of artist Richard Ankrom entitled *Freeway Driver*. Anthropometry — the study of body sizes within the population of users — plays a prominent role in an amusement park incident described in *The Perilous Plunge* and in the commercial aviation story, *Under the Radar*. Classic control, display, and workspace issues (knobs and dials psychology, as it is sometimes called) were of preeminent importance in the untimely death of musician John Denver, which is presented in *Rhymes and Reasons*.

Although not often viewed as a glamorous or particularly interesting topic, human factors in the maintenance of complex equipment and systems plays an increasingly important role in overall system reliability and safety. Maintenance procedures and system design are prominently featured in a number of chapters in the book, including *Caught on Tape*, the nightmarish tale of AeroPeru Flight 603. From the passenger's point of view, ergonomics in the maintenance hangar can be every bit as important as ergonomics in the cockpit.

The stories of medical error reflect procedural aspects of systems more than, perhaps, any other factor. In *Safer than Safe*, the rush to scale up and industrialize the highly complex, tabletop laboratory process for making Salk polio vaccine resulted, tragically, in the distribution of live polio vaccine during the initial days of the first inoculation efforts in 1955. Similarly, a rush to help a patient undergoing a harmless MRI scan has appalling consequences, as told in *Event Horizon.* And in *The Embryo Imbroglio* it is an unthinkable yet surprisingly simple deviation in procedure that results in a very big surprise for two couples undergoing fertility treatments. Viewed in retrospect and with the facts at hand, countermeasures to each of these types of accidents can be put into place so they are less likely to happen in the future.

Many stories in the book, however, like *The Atomic Chef* (a nuclear accident) and *End Game* (a ferry disaster) incorporate multiple causative factors and reflect systemic organizational problems. These *macroergonomic* issues include the organization's safety culture (or lack of one), organizational structure, process oversight and verification, performance monitoring, and personnel skill and training. Any contemporary discussion of human error must include these broader systems issues, especially in light of continuous expansion of system size, complexity, and interdependencies. In this sense, the field of human factors in design has progressed far beyond the "knobs and dials" focus of the past — as have the technologies on which the discipline is focused.

Although it is the final seconds of a disaster that capture the public's attention, the most riveting aspects of many accidents are the interactions between users and technology in the time leading up to the terminal event. It is the intent of each of the stories in this book to lay out the contributing factors in a manner that lets the reader understand how and why users did what they did and do what they do — without suffering the same, sometimes horrific consequences.

As in *Set Phasers on Stun* I have deliberately not analyzed each story after the fact and laid bare the contributing factors using the jargon of behavioral science. Rather, I have told each story simply yet carefully, with a sufficiently detailed description of the setting, the technology, and people involved, so you, the reader, can see and understand what happened. The facts are presented to find and consider. At the end of each story, ask how this system might have been designed, managed, and operated to avoid the unintended outcome. Remember that design-induced errors and the resulting accidents are not random events. If such were the case there would be no hope of improving system reliability by improving the user interface, which is most certainly not the case.

The following pages will take you to airports and airline cabins, an amusement park, a fertility clinic, a pharmaceutical plant, an emergency dispatch center, the Olympic games, and a bank; to hospitals, spacecraft, ships, and cars. From the coasts of Peru and Monterey, the Mediterranean Sea, in orbit aboard the International Space Station, the freeways of Southern California, the back roads of France, the battlefields of Afghanistan, and a nuclear plant in Japan — this is *The Atomic Chef.*

THE ATOMIC CHEF

Hisashi Ouchi tapped the shiny bottom of the upturned stainless steel bucket with the tips of his fingers to empty the contents one last time. A drop of bright yellow liquid dripped from the rim. He placed the bucket on the workbench and stepped back to inspect the result of the morning's efforts: the last of the partially filled glass beakers of uranyl nitrate, a concentrated solution of 18.8% enriched Uranium-235 — today's *theme ingredient.*

Some might think it odd to admire a purely industrial product such as the heavy fluid he had just poured from the bucket to the beaker, but to Hisashi it was visually attractive, particularly the way the transparent glass wrapped itself around the thick saffron solution they had brewed, giving it all the appearance of a crystal bowl of fine lacquer. But this, of course, was not a bowl of pretty yellow paint. It was the material for fabricating yellow cake — nuclear fuel pellets for Japan's high-tech Joyo ("everlasting sun") experimental fast breeder reactor. Once poured into the precipitation vessel with the previous batches, all 40 liters would be thoroughly mixed and, later, dried, powdered, and molded into stout little cylinders of uranium at another location. A few months from now the finished pellets would be stacked neatly inside metal tubes in the Joyo reactor at its next scheduled refueling, configured for sustained, controlled, and contained nuclear fission. It was regal material indeed, and cooking up the liquid uranyl nitrate was an impressive bit of chemistry, especially considering that it had all been done in their trusty stainless steel bucket. Chairman Kaga would be proud.

❖❖❖

Hisashi and his coworkers, Masato Shinohara and Yutaka Yokokawa, had completed the same steps of chemistry seven times since yesterday afternoon, each time producing 5.7 liters of fluid in their metal pail. They began by carefully weighing out 2.4 kg of the uranium powder, technically known as triuranium octoxide, and adding it to the empty bucket. Next, they measured and poured in a liter of pure water and mixed it thoroughly into a soupy paste. The 6.49 liters of nitric acid came next, which was measured precisely and added to the triuranium octoxide and water mixture, a little at first and then more as the uranium paste blended with the acid. They mixed and stirred, this way and that way, around and around and back and forth, until all that remained was the smooth paintlike solution with no clumps or lumps.

As with the preceding batches, transferring the mixture into the smaller beaker from the bucket would make it easier to pour this last batch of uranyl nitrate into the slightly elevated and inaccessible precipitation vessel. The partially filled beaker was smaller and lighter than the bucket. He had to hold it up high and pour the liquid into the funnel his coworker held over the precipitation vessel's small round opening. They had gone to so much trouble to make sure that all of their measures were exact, and they certainly didn't want to spill any of the U-235 solution onto the floor or the equipment.

Each bucket of concentrated uranyl nitrate had taken the three-man crew about an hour to produce. This did not include the many days of preparation here in the conversion building at the northwest corner of the JCO Nuclear Fuel Processing Facility in Tokaimura, a small village on the Pacific coast northeast of Tokyo. They had made four batches in the afternoon the day before. And now, 10:30 the next morning, the 30th of September, 1999, the precipitation vessel held six buckets of material — four from yesterday and two from today. Only the last bit of the last batch, which now sat on the workbench beneath the admiring eyes of Hisashi Ouchi, had not been added. The electrically powered mixing blades inside the big bowl-shaped precipitation vessel turned slowly, awaiting the final beaker of blended ingredients. Masato would return in a minute to help Hisashi get the last of this batch into the funnel and the precipitation vessel.

❖❖❖

Hisashi, 35 years of age, had spent much time at the JCO plant during his 16 years with the company, but he had not worked exclusively in the conversion building, at least not until recently. The company had gone through some financial difficulties this past year, and, with the layoff of nearly a third of their people, he and the others who remained had been asked to take on more responsibility than in the past, including the campaign to produce the enriched U-235 for the Joyo reactor. They had had no special training for the new line of work — or much of any training at all in the physics of fission. Yet Hisashi, Masato, and Yutaka had accomplished a great deal in these few short days, certainly more than they would have accomplished with a larger crew or if forced to use all of the built-in processing machinery and the accompanying slow procedures. There was some truth to that old adage: *Give a small group of capable men good and simple tools and they can do just about anything.*

❖❖❖

An enriched uranium solution of 141 kg was first produced at the facility in 1993, when JCO Co., Ltd. was known as Japan Nuclear Fuel Conversion Company. A campaign to manufacture a batch of triuranium octoxide was undertaken in 1994 and was followed by two campaigns for batches of uranium dioxide in 1995 and 1996. With a handful of other campaigns running up until June of 1998, the facility produced a total of 963 kg of enriched uranium-based material, all for the Joyo experimental fast breeder reactor. Unlike the more automated and controlled manufacturing processes used to make standard uranium fuel for commercial reactors, the process here in the conversion building was very hands-on. It was also a "wet" as opposed to a "dry" process. The requirements of their customers and their unique nuclear reactors demanded a special operation and a custom product.

Most of the time Hisashi and the other JCO employees at the Tokai Works (as the facility was sometimes called) were involved in the production of uranium dioxide (UO_2) from enriched uranium hexaflouride (UF_6), the resulting substance being a relatively common 3 to 5 percent enriched uranium. The fuel for the Joyo breeder reactor required more of a punch, so the manufacturing process, as well as the enrichment level of the resulting product, was different this time around. Most importantly, the main ingredient — uranium powder — was much more concentrated. The uranium fuel they were now manufacturing was to be enriched to 18.8 percent, only slightly less than the 20 percent maximum allowed by JCO's operating license.

The single-story conversion building where the operation was taking place was made of cinder blocks. The structure was rather small, only 15 or so meters on a side. Three interconnected rooms lay inside the square outer walls. A short hallway linked the largest room to an adjoining building. The smallest room, a narrow L-shaped space, contained most of the processing equipment and the precipitation vessel which now held the previous batches of concentrated liquid uranium. Hisashi and the others paid close attention when working around the unusual equipment in the tight corridors.

The interior of the conversion building was a blend between an industrial chemistry lab and a small factory, with an odd mixture of different shapes and sizes of tanks, pipes, pumps, and support structures. A dissolving process performed in the dissolution tank where raw uranium oxide (U3O8 powder) was mixed with nitric acid was the starting point. Pipes from the dissolution tank ran to a solvent extraction system, consisting of a pump, a solvent extraction column, and an extraction stripping column. Both the solvent extraction column and the extraction stripping column were tall and narrow.

Exiting the stripping column, the resulting solution flowed through a pump and more plumbing to two tall and narrow buffer columns, and, lastly, to the precipitation vessel, a bowl-shaped tub about 45 cm wide and 60 cm deep. An internal mixer in the precipitation vessel stirred the contents. Mixing materials in the precipitation vessel could generate heat from common chemical reactions, so the tank was fitted with an integral cooling system — a 2.5-cm-thick metal water-filled jacket that surrounded the bottom half of the tank. Pipes for the cooling water in this closed-loop cooling system ran through the cinder block wall of the conversion building to a pump, drain valve, and heat exchanger, all of which were located outside on the other side of the cinder block wall.

At this stage of the process the contents of the precipitation vessel were infused with gaseous ammonia to create ammonium diuranate. The ammonium diuranate was filtered as it exited the bottom of the precipitation vessel and transferred to large flat trays. The trays, in turn, were to be loaded into a furnace, and converted into triuranium octoxide. They were later transferred to another furnace with an ammonia-containing cover gas and converted to purified uranium dioxide, the desired end product of the operation.

The entire system and process — starting with the mixing of the powdered uranium oxide and nitric acid in the dissolution tank and ending with the purified uranium oxide removed from the second furnace at the other end — had been approved by JCO when the process was designed. The process had also been approved by the Japan Science and Technology Agency, which concluded that there was "no possibility of criticality accident occurrence due to malfunction and other failures." The government regulators had determined that the process was foolproof and that there was no chance of things going terribly wrong.

❖❖❖

The irony of the moment would not have been lost on anyone familiar with the facilities in the conversion building, the proper handling of fissionable material, and the decidedly low-tech production methods being employed by the well-intentioned three-man crew. Here sat a small fortune in laboratory and process equipment, designed to produce liquid uranyl nitrate (18.8 percent enriched uranium-235) in carefully controlled and metered batches. The maze of metal that lay between the solution tank at the beginning of the process and the entrance to the precipitation vessel about halfway through was designed to manufacture precisely the same material the three-man crew had mixed up in their bucket. The complexity and uniqueness of the hardware alone would make most anyone believe it not possible to replicate such a process in a bucket, but this was not the case. Unbeknown to the Japan Science and

Technology Agency, which was responsible for overseeing the operations at the Tokai Works, JCO had developed procedures to bypass the equipment, saving considerable manufacturing time, cost, and labor required to clean out all of the equipment after a batch had been made.

The first change in the procedure was developed shortly after the system went on-line years before and involved bypassing the initial dissolution process in the dissolution tank. Instead of mixing the raw uranium oxide, water, and nitric acid in smaller batches in the dissolution tank at the beginning of the process, JCO personnel had developed a procedure for mixing the ingredients in the stainless steel bucket. The change was circulated around the company, and formal approval was given by the manufacturing division and the division of quality assurance. The safety management division was not consulted about the change. A formal procedure was printed for use by the conversion building workers and used for a number of years. Contrary to the original licensing agreement for the operation, JCO did not notify any regulatory body of the change. Nor did they submit the change to a special criticality review committee which was to have been created under the original terms of their licensing. In fact, no such committee had ever been formed.

As months passed, it was realized that using the bucket saved so much time that another simplification was developed and approved: batches of enriched uranyl nitrate were now being pumped and accumulated in the tall narrow buffer columns located a few feet beyond the dissolution tank. As with the first change in the process, no individual, department, or regulatory agency with a full understanding of the process and the reasons behind it was consulted.

Unfortunately for Hisashi, Masato, and Yutaka, but also much of the population of the entire prefecture, JCO's willful modification of the process and the regulatory agencies' failure to monitor the operation would have dire consequences. The enriched uranium manufacturing system laid out before Hisashi was designed to make small batches of uranium solution, each containing no more than 2.4 kg of raw uranium — the amount Hisashi had added to *each* of the seven bucketfuls that had been mixed yesterday and today. The designers of the original process and the government regulators who approved it had no idea that anyone would try to make larger batches or circumvent part or all of the process.

A third and more critical modification had been made to the process in the preceding days, the result of the crew's well-intentioned efforts to speed up the process further, thus saving the company even more money. Why, Hisashi and the others had wondered, spend all this time making little individual batches and pumping them into the tall buffering tank and individually into and then out of the precipitation vessel? It seemed like such a waste of time. Why not simply mix each of seven 5.7-liter batches in their bucket and mix them all together in the large precipitation vessel which had more than twice the volume necessary to contain the entire 40 liters of concentrated enriched uranium liquid? Necessity, as the saying goes, is the mother of invention.

❖❖❖

So this is what the crew had done: bypassed the dissolution tank, the solvent extraction and stripping columns, and the two buffer columns by mixing the material in their bucket and pouring it directly into the precipitation vessel with the aid of a glass beaker and a funnel. They had concocted one big yellow frothy pot of enriched uranium solution. Neither the three-man crew nor the other involved employees and managers at the Tokai Works had any idea that their sizable cauldron of U-235 was teetering on the precipice between a state of rest and an avalanche of neutrons, the near-instant result of which would be a full-blown and sustained nuclear chain reaction. Unlike the tall and narrow extraction and buffer columns, each of which was designed to prohibit the accumulation of sufficient mass to produce a chain reaction, the bowl-like precipitation vessel was geometrically favorable to such an event by keeping the contents very compact and close together. And although it was not intended to hold such a large volume of highly enriched uranium in the normal small-batch processing mode, the precipitation vessel was perfectly capable of doing so. Furthermore, it was lined with the 2.5-cm-thick blanket of water to remove excess heat during conventional chemical reactions. The lining of water also served as a reflector, however, bouncing neutrons back into the nuclear brew and helping to sustain a nuclear chain reaction should it begin.

❖❖❖

Masato Shinohara returned to the small room moments after Hisashi Ouchi placed the empty stainless steel bucket on the workbench, abruptly ending Hisashi's private thoughts about the meaningfulness of their efforts and the beauty of the product they had created. Yes, it was time to get back to work. Yutaka Yokokawa was obviously staying in the adjoining corridor on the other side of the wall to attend to related things. Anyway, only two people were needed to pour the remaining batch.

Hisashi and Masato bounced a series of simple questions and responses back and forth, each asking if the other was ready to finish up the job. They both stepped up to the elevated precipitation vessel, one holding the funnel and the other holding the large glass beaker of enriched U-235.

Up went the funnel, its small end inserted through the 25-

cm-wide round opening in the top of the bowl-shaped tank, the funnel's upturned wide end waiting for the fluid.

Up went the glass beaker of uranyl nitrate liquid.

A stream of attractive yellow liquid flowed off the lip of the glass beaker and into the funnel, down the spout and down into the precipitation vessel where the slowly rotating blades blended it into the nearly 40 liters of enriched uranium lying in wait and waiting for life.

The scale of misfortune tipped against them within seconds as the total mass of enriched uranium exceeded the limits defined by the laws of nuclear physics.

Liquid uranyl nitrate foamed up lightly like boiling yellow milk. Intense blue light burst from the precipitation vessel, filling the room in a misty aqua glow. A neutron detector at the Japan Atomic Energy Research Institute a full 2 kilometers away recorded a massive spike in neutron radiation and would continue to do so for the next 20 hours. Nuclear criticality had been achieved.

An intense and strange heat washed over Hisashi. He stumbled back, stunned, and slumped halfway to the ground. Instantly, it seemed, he was nauseated, nauseated like he had never been before. To his side was Masato, looking numb and strangely ill, yet only seconds had passed. The room was blue, yes blue! This was incredible. Nothing made sense. They had been pouring the uranyl nitrate into the precipitation vessel. That was all. Why would there have been a fire? But it didn't seem quite like a fire. The heat was strange. It went straight through him, as if he were invisible. But it had to be a fire. He felt so hot.

His stomach turned. Oh, it was going to happen. He could not stop it. The vomit erupted up from his stomach, out and onto the floor. It came and came again. Then it stopped for a few moments, but the nausea flowed back through him once more. This time his stomach turned low and deep and he knew he had lost complete control.

Hisashi could no longer stay on his hands and knees and collapsed further, lying down on the cold floor now, low and prone. But the blood still seemed to drain from his brain and the color from his sight.

❖❖❖

The paramedics from the local fire department eventually came (although there was not a fire) and wrapped each of the three men in clear plastic and wheeled them off on gurneys. They were rushed to the National Hospital in Mito in an ambulance, then helicoptered to the National Institute of Radiological Sciences in Chiba Prefecture.

Back at the Tokai Works, efforts to stop the uncontrolled reaction were hampered by lack of information, plans, and preparedness for an event that "was not possible." The nuclear chain reaction was halted 20 hours later only after a crew crushed the pipes leading from the blanket surrounding the precipitation vessel, thereby draining the water onto the ground outside the conversion building. Removing the reflective water was just enough to bring fissioning to an end. In the end, 436 JCO employees, nearby residents, and workers had been exposed to radiation from the event. Economic losses, including financial costs to regional business and agriculture stigmatized by the accident, eventually exceeded 6 billion yen.

❖❖❖

Hisashi Ouchi was subsequently transferred to Tokyo University Hospital on October 2 where he received blood transfusions, large-scale skin transplantation, and hematogenous function recovery treatments. His injuries were massive: burns, organ damage, fever, decreased immunocompetence, a high white cell count, respiratory deficiency, and continued disturbance of consciousness. Calculations showed he had received 18 sieverts of penetrating radiation, roughly 9,000 times the typical annual exposure and more than three times the suspected fatal dose. In an effort to restore his lymphatic cells and blood-generating capability, he received stem cells from his brother in a first-ever procedure for radiation victims.

Masato Shinohara, who was in a slightly different position next to the precipitation vessel, received 10 sieverts of penetrating radiation, more than 5,000 times the typical annual exposure and two times the usual fatal dose. A suitable adult donor could not be obtained, so he received a transfusion of blood stem cells drawn from a newborn's umbilical cord. Yutaka, who had been in the next room, received a dose of 3 sieverts.

Hisashi Ouchi died of multiple organ failure on December 21 at Tokyo University Hospital. Masato Shinohara, who initially regained his strength and showed hopeful signs of recovery, died seven months later on April 27 at the Institute of Medical Science, University of Tokyo. Yutaka Yokokawa, shielded by the wall and his distance from the precipitation vessel, survived. JCO's fuel processing license was revoked by the Japanese government on March 29.

REFERENCES AND NOTES

Criticality accident at Tokai nuclear fuel plant, Japan (2000). Amsterdam: World Information Service on Energy, October 15.

Dolley, S. (1999). *Japan's nuclear criticality accident.* Washington,

D.C.: Nuclear Control Institute.

Japan takes stock after Tokaimura nuclear accident (1999). *CNN.com www site*, October 2.

Japan's nuclear criticality accident at a fuel conversion facility, December 9 technical summary (1999). Japan Atomic Industrial Forum, Inc.

Japan's nuclear regulator says sorry for Tokaimura accident (2000). *Environment News Service www site*, July 7.

Lamar, J. (1999). Japan's worst nuclear accident leaves two fighting for life. *British Medical Journal*, 319, October 9, 937.

Lamar, J. (1999). Japan widens its investigation after nuclear accident. *British Medical Journal*, 319, November 20, 1323.

Landers, P. (1999). Diary of nuclear accident: Japan wasn't ready. *The Wall Street Journal*, October 8, A17

McCoy, F. R., McLaughlin, T. P., and Lewis, L. C. (1999). *Trip report of visit to Tokyo and Tokai-mura, Japan for information exchange with Government of Japan concerning the September 30, 1999 Tokai-mura criticality accident.* Washington, D.C.: U.S. Department of Energy.

Media advisory: technical briefing on the radiation accident in Japan (1999). Press Center, International Atomic Energy Agency, October 1.

Meshkati, N. and Deato, J. (2000). Japan must commence nuclear reforms. *The Japan Times*, October 2.

Normile, D. (1999). Special treatment set for radiation victim. *Science*, 286, October 8, 207-209.

Normile, D. (2000). Exposure levels tracked around nuclear accident. *Science*, 288, May 19, 1153.

NRC review of the Tokai-mura criticality accident (2000). Washington, D.C.: Division of Fuel Cycle Safety and Safeguards, Office of Nuclear Material Safety and Safeguards, U.S. Nuclear Regulatory Commission.

Nuclear issues briefing paper no. 52; Tokai-mura criticality accident (2000). Melbourne, Australia: Uranium Information Center

Nuclear safety rules ignored, panel says (1999). *Los Angeles Times*, December 19, A14.

Plant in Japanese nuclear accident has license revoked (2000). *Wall Street Journal*, March 29, A3.

Report on the preliminary fact finding mission following the accident at the Nuclear Fuel Processing Facility in Tokai-mura, Japan (1999). Vienna: International Atomic Energy Agency.

A summary of the report of the Criticality Accident Investigation Committee: January 19, 2000 revision of December 24, 1999 report (1999). Nuclear Safety Commission, Science and Technology Agency of Japan.

THE EMBRYO IMBROGLIO

Embryologist Michael Obasaju, PhD, peered into his microscope at the assortment of human embryos before him in the petri dishes on the hot plate, each embryo only three days old, each the beginning of a human life. People were often surprised to learn that embryos such as these were not all identical in appearance at this early stage. In fact, each looked unique in some way. A few were dividing rapidly while others were slower, less aggressive. Some were of uniform shape and a few appeared fragmented and irregular. The cytoplasm within the cells varied also — from clear and transparent to clou dy and mirky. These differences were not signs that one embryo would develop into a tall person, a short person, a black person, or a white person; the differences were only a suggestion of the health of the embryo and the likelihood of it taking root once placed inside the uterus of the woman seeking pregnancy. Like picking the best-looking fruit on the tree, his task at the moment was to examine each one carefully and separate them — paying particular attention to finding the clearest, most uniform, and fastest-growing little clusters of cells, the embryos having the highest probability of survival once placed in the patient.

Two women were to undergo in vitro fertilization in the IVF New York clinic that day, at 230 Central Park South in Manhattan, under the care of Dr. Lillian Nash, age 71, an infertility specialist. As Dr. Nash's embryologist, it was Dr. Obasaju's job to sort the embryos and prepare them for implantation. Three days before, on April 21, he had assisted Dr. Nash when she removed eggs from the same two patients. Each woman had undergone daily hormone injections for the week leading up to the 21st, and on that day Dr. Nash surgically harvested a number of eggs from each. Dr. Obasaju had introduced each woman's eggs to the sample from her husband and placed them in the petri dishes with the growth culture. Nature ran its course in the incubator, set to body temperature, and the result today, three days later, was a healthy clutch of living embryos before him in the culture dishes on the hot plate at his workstation.

Dr. Obasaju had a nice assortment of seven embryos for Mrs. Donna Fasano from Staten Island, the first woman scheduled for in vitro fertilization. They often liked to have more embryos than this so they could select only the very best from the batch later on. Sometimes there was a problem finding high-grade embryos, especially from older women nearing the end of their reproductive years. But this was not the case here; all seven specimens looked very good, and only four were needed for implantation. So he separated four and gently suctioned them up and placed them in a separate dish. They would be placed carefully inside Mrs. Fasano's uterus later by Dr. Nash with the assistance of Dr. Obasaju.

Like anyone working in the fertility business, Dr. Obasaju knew that in vitro fertilization was very much a game of chance and that the process involved calculating and manipulating probabilities to maximize the chance of pregnancy and a successful birth. The things not under their control were the age of the woman and the quality of the embryos, although they certainly had say over which patients they would take on and which of the available embryos would be used. They also had control over the number of embryos to be implanted, the techniques for creating the embryos, and the procedures for implanting the embryos. Each decision or method along the way could raise or lower the probability of the patient getting pregnant.

The probability of a single embryo resulting in pregnancy was somewhere between 0 and 1, but usually toward the low end of the scale, especially when the embryo was of poor quality. The probability of pregnancy could be increased, however, by implanting additional embryos with the hope that one of them would take hold. But with multiple embryos in the woman's womb there was the off chance that more than one embryo — or even all of the embryos — might develop, thereby resulting in twins, triplets, quadruplets, and so on depending on how many embryos were implanted. In the case of Mrs. Fasano they had settled on four embryos, a balance between the probability of pregnancy and the chance and acceptability of multiple births.

❖❖❖

The second woman expected at the clinic that day for implantation of her embryos was Mrs. Deborah Perry-Rogers, age 33, a nurse from Teaneck, New Jersey. Dr. Nash was able to harvest 25 eggs from Mrs. Perry-Rogers on the 21st, and 20 of them had been fertilized in the incubator. Dr. Obasaju studied each of the 20 embryos carefully, sorting them by their size, symmetry, and clarity. He gently suctioned up the six best-looking embryos and placed them in a dish for implantation into Mrs. Perry-Rogers later that day. An additional 10 embryos also looked good, and he suctioned them up and placed them in a separate dish. They would be taken to another lab later and frozen for possible use another time should this implantation trial with Mrs. Perry-Rogers be unsuccessful, and storing the embryos in the freezer would be less stressful on the patient than going through the egg harvesting regimen and surgery again. This left four low-grade embryos, embryos that were of such low quality that they would have a very low probability of implanting themselves in the uterus and developing into a healthy fetus. These four embryos would be discarded, and he suctioned them up and put them in their own petri dish on the hot plate along with the other dishes containing the sorted embryos from Deborah Perry-Rogers and Donna Fasano.

❖❖❖

Each patient arrived at the IVF New York clinic shortly thereafter for in vitro fertilization and a new chance at pregnancy. Donna Fasano's and Deborah Perry-Rogers' appointment times were different, however, and they did not cross paths.

Back at his workstation, Michael Obasaju organized his instruments and his large assortment of petri dishes holding the human embryos. Each dish had been labeled, but they were all the same size and shape and one had to keep the names of these patients straight. Dr. Nash was talking with Donna Fasano of Staten Island, now lying on the table on her back with her feet in the foot rests. No doubt about it, it was an undignified position in which to put someone, but necessary for the procedure. The actual insertion would take only a moment and feel no different than a normal Pap smear. Unlike the egg harvesting procedure a few days back, there was no need to do any cutting or stitching, no need for anesthesia. They would put the embryos into a thin strawlike device, slip it through her cervix, and squeeze the embryos into her uterus. All Mrs. Fasano had to do was relax and think positive thoughts about the prospects of becoming a mother after all the years of frustration.

Michael Obasaju had served as Dr. Nash's embryologist for

quite some time and had learned to synchronize his own work with hers. He did not want Dr. Nash or the patient to have to wait for him, just as he did not want to have the embryos loaded in the insertion instrument too far ahead of time. But now seemed like a good time to move on, so he prepared a drop of culture medium and then carefully suctioned up all four embryos from a petri dish on the hot plate. The embryos — about the size of a small pin head — were just visible to the naked eye if you got up close and looked very, very carefully under good light. The long and thin embryo transfer catheter had a syringe on one end, and he pulled it back to draw the tiny amount of culture medium and the four suspended embryos into the tube.

The patient and Dr. Nash were waiting and ready to get on with the procedure, but just prior to turning to hand the embryo transfer catheter to Dr. Nash, Michael Obasaju had one of those odd, fleeting feelings that something was amiss, that something had been left undone, that he was standing at the wrong track in the train station. He paused for a moment and looked back at his considerable assortment of petri dishes. Yes, there were the three dishes for Mrs. Perry-Rogers: one dish containing the six good embryos to be implanted, another holding the 10 embryos to be frozen, and the third for the four low-grade embryos to be discarded. And there were the dishes for Mrs. Fasano: one dish containing the three embryos which were not going to be used and another dish holding the four embryos to be implanted. The last dish still contained four embryos, and at that moment Dr. Obasaju recognized his most recent error: he had mistakenly loaded the embryo transfer catheter with the four lower-grade embryos from Mrs. Perry-Rogers, the embryos that were to be discarded. But it was Mrs. Fasano — not Mrs. Perry-Rogers — who was lying on the table and waiting patiently for her next chance at pregnancy and the blissful state of motherhood.

The clock was ticking and Dr. Obasaju thought about what to do. He had obviously been busy the past few minutes, and starting over with a new embryo transfer catheter or cleaning out the one he had just loaded might look unusual or arouse concern in the others in the room. On the other hand, the four embryos now in the embryo transfer catheter were the four very low-grade embryos from Mrs. Perry-Rogers. True, they were not produced by the woman on the table, but they were of such low quality that the odds of any one of them successfully implanting in the uterus seemed miniscule. He, after all, was the specialist who had classified the embryos and knew just how poor these four little embryos looked. And there was also the matter of the overall odds of pregnancy from the in vitro procedure, odds which were not especially high at best and greatly different from one clinic to the next. For the U.S. as a whole, 37 percent of all in vitro procedures for women under 35 years of age had resulted in live births the year before, in 1998. Yet Dr. Nash's success rate was 13 percent, despite the fact that she usually implanted more embryos than did most clinics during each in vitro trial. It was difficult to argue with official U.S. government statistics and the CDC which kept track of all these things.

Dr. Obasaju chose his course of action and removed the lid from the dish containing Mrs. Fasano's four embryos. He quickly prepared another batch of culture medium, suspended the four embryos in it, and sucked them into the transfer catheter. His mistake, he reasoned, would be forgotten and no one would ever know the difference. He handed the embryo transfer catheter containing four embryos from Mrs. Deborah Perry-Rogers and the four embryos from Mrs. Donna Fasano to Dr. Lillian Nash, who gently inserted the thin catheter through the opening in Mrs. Fasano's cervix and squeezed the syringe.

❖❖❖

The next few weeks gave Michael Obasaju more than enough time to think about that day and what he might have done differently. Everything seemed to go just fine with Mrs. Perry-Rogers; he had prepared her six good embryos and Dr. Nash had implanted them. It would be a good thing if she became pregnant. The real issue was the in vitro fertilization procedure with Mrs. Fasano. Yes, this is what gnawed at him. It had been a stupid, careless mistake, one he could have avoided had he never had the two women's embryos anywhere near each other and if he had followed the clinic's written procedures that had been approved by the New York State Health Department. And his decision to say nothing about his error and prepare the second batch of embryos for implantation into Mrs. Fasano was indeed the wrong one, but he reminded himself that the odds of pregnancy were really quite low given the clinic's track record, and especially low for Mrs. Perry-Rogers' embryos considering their poor quality. But then again, unfortunately, there was the outside chance that Mrs. Fasano might get pregnant, not only with one of her own embryos, but with one of Mrs. Perry-Rogers' embryos, or, heaven forbid, with one of each. She could be pregnant with two babies who were genetically unrelated to each other! Even this last outcome might not be so seemingly cataclysmic and might even go unnoticed were it not for the fact that Mrs. Donna Fasano and her husband from Staten Island were white and Mrs. Deborah Perry-Rogers and her husband from Teaneck, New Jersey were black.

❖❖❖

The results were in by the middle of May. As far as these two infertility patients were concerned, the Obasaju and Nash team had batted 500. By another measure, though, they had slammed the ball out of the park with a runner on base.

The downside was that the in vitro fertilization trial with Mrs. Perry-Rogers was not successful and she was not pregnant. However, Mrs. Fasano was pregnant and she and her husband were understandably overjoyed. Dr. Nash relayed the good news to Dr. Obasaju who, on May 20, after a little more time to think things over, informed Dr. Nash that there was an important matter they needed to discuss privately. It concerned the events of April 24 and the in vitro procedures with Donna Fasano and Deborah Perry-Rogers. He had made an error, he confessed to her, and explained in detail what had happened on that day and the possible consequences. There was the possibility, however remote, that Mrs. Fasano might be carrying another couple's child.

Dr. Nash immediately relayed the information to the Fasanos, now a number of joyful weeks into their long-sought pregnancy. Upset and incredulous, they went to an independent lab for in utero DNA tests. The results were difficult to fathom. Donna was pregnant with two fetuses, not one. One child was the genetic offspring of Donna and her husband; the other fetus was entirely unrelated to them. Dr. Nash had told the Fasanos nothing about the source of the other embryos implanted in her. But after talking things over at great length they decided there was only one course of action. What else was there to do but move on with the pregnancy and hope for healthy babies? Nothing, really. So Donna Fasano courageously put her worries aside as best she could and enjoyed her pregnancy with her "gestational twin siblings."

Upon hearing the news from the DNA tests Dr. Nash felt it prudent to inform Deborah Perry-Rogers and her husband on May 28 that one of their embryos had, after all, successfully implanted, and that they were going to have a baby. There was a slight problem, obviously, considering that Deborah Perry-Rogers was not pregnant. There had been an error during the in vitro procedure with another patient, and some of Deborah's embryos had been used by mistake. The Rogerses, no doubt, had as difficult a time as the Fasanos accepting what they were told. Their child was growing inside a Mrs. Donna Fasano, a white woman from Staten Island!

❖❖❖

On December 29, 1998, Donna Fasano gave birth to two healthy boys: one white, of European descent, and one black, of African descent. The Fasanos named the white infant — their own biological son — Vincent, and they choose the name Joseph for his "gestational brother," the biological offspring of Deborah Perry-Rogers and her husband Robert Rogers. For the next months the Fasanos doted over the boys as would any loving parents. Vincent and Joseph ate together, bathed together, slept in the same crib, and were rocked in the same swing, certainly unaware of the building tempest over their disposition and future.

The flood gates opened on March 16, 1999, when the Rogerses filed a lawsuit against the Fasanos and the involved medical personnel, seeking custody of their biological son, Joseph, and claiming negligence on the part of Drs. Nash and Obasaju. Deborah Perry-Rogers became "emotionally scarred and overcome with depression," the filing said, and the Fasanos had "adopted a hostile stance" toward the Rogerses, the latter statement being a bit of a stretch considering that the Fasanos did not know the identity of the Rogerses and had severed all ties with Dr. Nash. So it was on this day that the Fasanos learned the identity of Joseph's biological parents, and it was also on this day that the whole affair became painfully public. The following day on March 29 the New York State Health Department began an investigation into the entire incident and

IVF New York. The press camped out on the Fasanos' front lawn, pestering them with requests to be interviewed, the photographers vying to get the first picture of the two boys. But the Fasanos insisted on their privacy, closed their blinds, and refused to parade the boys before the media. Everyone — the Fasanos, the Rogerses, and their sons — were unsuspecting victims of the procedures and the decisions of Dr. Michael Obasaju and the supervision he received by his superiors.

The next day, on March 30, in a reasoned response, the Fasanos announced through their lawyer that if DNA tests confirmed that Deborah Perry-Rogers and her husband Robert Rogers were the biological parents of Joseph, then they, the Fasanos, would relinquish custody of the baby boy. "We are giving him up because we love him," said Donna Fasano. It was not an easy decision, but it seemed to be the right thing to do considering everything that had happened. At the same time, however, Donna Fasano felt strongly that Vincent and Joseph had become emotionally attached to one another, just as she had become attached to Joseph, naturally. She wanted them to remain close as they grew up. Her affection toward both boys grew over the ensuing weeks. The DNA test results eventually confirmed what everyone knew, and on May 10, when the boys were over five months old, the Rogerses met the Fasanos at the Fasanos' home on Staten Island to take custody of their son. It could be said that things did not go well.

The Fasanos' lawyer had drawn up a "visitation agreement" as part of the custody accord, stipulating that Joseph be delivered to the Fasanos' home on Staten Island for regular weekend visits with them and Vincent throughout his childhood. The Rogerses would be penalized $200,000 if they did not live up to the terms, according to the papers. It was, as the Rogerses' lawyer said, "unreasonable," "onerous," and "one-sided." But here was their son in their arms, their very own son after all these painful months, and it was so hard to believe — but it was true. So they signed the visitation agreement and took him across the river to New Jersey where he would have a new home and a new name, Akeil Richard Rogers.

❖❖❖

If only the ending was this simple. The Fasanos were due their own lawsuit, which they filed in the Manhattan State Supreme Court, stating that "negligence, carelessness and malpractice of..." Drs. Nash and Obasaju caused the wrong embryo to be implanted into Donna Fasano. Further, Dr. Nash was charged with violating confidentiality rules by having divulged the identity of the Fasanos to the Rogerses during Donna Fasano's pregnancy.

Also, it soon became apparent to everyone involved that the visitation agreement was not working out. Initially, at least, the Rogerses delivered Joseph — now Akeil — to the Fasanos' residence in Staten Island for the first weekend visit as stipulated by the papers they had signed. It was a long trip, and the Fasanos and their son Vincent were under no obligation to reciprocate by visiting the Rogerses in New Jersey. Things really went astray during a subsequent get-together when Mrs. Fasano insisted on calling Akeil "Joseph" and referring to herself as "mommy." The Rogerses could not tolerate this, said their attorney. "We will fight that to the end." But the Fasanos' lawyer was equally adamant: "The boys were brought into life by Mrs. Fasano. Whatever legal terms you want to say, she'll always in her heart feel she is the mother." The Rogerses stopped delivering Akeil to the Fasanos for visits with Vincent, and on June 24, when the boys were six months old, the Fasanos sued the Rogerses "to force the child's black genetic parents to live up to... the... visitation agreement."

❖❖❖

Arguing before a justice of the New York State Supreme Court in Manhattan, Deborah Fasano and her attorney contended that Joseph should be considered both her son and the son of Mrs. Perry-Rogers. Justice Diane Lebedeff ruled on July 16 that Joseph belonged in the permanent custody of his biological parents, Deborah Perry-Rogers and Robert Rogers, although an arrangement for visits over the next two months in a neutral location in New Jersey would have to be made by the couples. Meanwhile, the attorney for Dr. Obasaju argued for the dismissal of any claims of emotional pain and suffering against him because the act of implanting the wrong embryos had not caused the actual injury. The wrangling over custody and visitation was a matter involving the two couples, he said, not a basis for a claim against Dr. Obasaju. Justice Lebedeff disagreed, stating "I can't purge my mind of the other circumstances of this case," and refused to drop the case against Dr. Obasaju at that time. The State Health Department faulted Dr. Obasaju for numerous procedural violations in the handling of the embryos, including not keeping embryos from different couples physically separated, and Dr. Nash for failing to oversee and monitor the actions of her embryologist. Dr. Obasaju was excused from practicing embryology at IVF New York.

REFERENCES AND NOTES

1998 assisted reproductive technology success rates, national summary and fertility clinic reports, for Lillian D. Nash, M.D., New York, New York (1998). Atlanta: CDC National Center for Chronic Disease Prevention and Health Promotion Reproductive Health Information Source.

1998 assisted reproductive technology success rates, national summary and fertility clinic reports, 1998 national summary (1998). Atlanta: CDC National Center for Chronic Disease Prevention and Health Promotion Reproductive Health Information Source.

Couple in embryo mix-up who gave up child sue to enforce visitation pact (1999). *The New York Times,* June 25.

Diskin, C. (1999). Test-tube mix-up spurs twin court proceedings. *Bergen Record Online,* September 25.

Embryo transfer procedure for in vitro fertilization (2001). *Advanced Fertility Center of Chicago, IVF Specialist Clinic www site.*

Grunwald, M. (1999). Fertilization flop: white baby has black 'twin'. *The Washington Post,* March 31.

IVF procedure: embryo replacement [embryo transfer] (2001). *IVF-infertility.co.uk www site.*

Maull, S. (1999). White couple gives infant to his black genetic parents. *The Topeka Capital-Journal,* March 31.

Maull, S. (1999). Couple sues doctors over misplaced embryos. *The News-Times Interactive Edition,* April 7.

Mom to give up 'twin' born in black-white embryo mix-up. (1999). *The Des Moines Register,* April 1.

New Jersey couple sue over an embryo mix-up at doctor's office (1999). *The New York Times,* March 28.

Report: doctor not alerted in misplaced embryo case (1999). *The News-Times Interactive Edition*, April 17.

Rohde, D. (1999). Biological parents win in implant case. *The New York Times*, July 17.

Yardley, J. (1999). After embryo mix-up, couple say they will give up a baby. *The New York Times*, March 30.

Yardley, J. (1999). Health officials investigating to determine how woman got the embryo of another. *The New York Times*, March 31.

Yardley, J. (1999). Birth mix-up avoidable, inquiry finds. *The New York Times*, April 17.

SIGNAL DETECTION

The long line of passengers waiting uncomfortably at the American Airlines check-in counter at Charles de Gaulle International Airport was *typical* of what one saw every day of the week at 8:40 a.m., two hours before the flight to Miami International Airport in Florida, a scant four months after 9/11. *Typical,* it should be pointed out, with one glaring exception: the lone passenger who had only now worked his way to the front of the queue and was sauntering up to the next available agent at the counter. This man was difficult *not* to notice. He was tall, about 6 feet 4 inches, of undetermined race, with a long ponytail, scruffy beard, and unkempt appearance. A small knapsack hung over his shoulder and he handed his ticket and passport over the countertop to the agent at the check-in station. He was on his way to Antigua in the Caribbean, via Miami, to visit relatives. The ticket was for a round trip. The passport was British. The agent opened it and typed the passenger's name at the keyboard. A stench snaked up over the counter.

One had to handle these matters with considerable sensitivity. When the agent's display rebounded with a glaring red code, no clue about the computerized CAPPS (Computer Assisted Passenger Profiling) warning was revealed to the odoriferous customer. It was the red, not green, color code that signaled a suspect passenger, a passenger targeted for follow-up screening by airline security. Sometimes passengers were selected at random by the system in order to keep inspectors on their toes and would-be terrorists slightly off balance, but a random selection appeared unlikely in this case. A set of conditions had been triggered in a program, a small number of the 40-odd undisclosed cues established by the U.S. Federal Aviation Administration. Perhaps it was the fact that the ticket had been purchased just four days before, a ticket bought with cash from an office in Paris, or the fact that the passenger had no verifiable address and no luggage to speak of. This was not a random selectee. The agent politely directed him to step over to the side a short distance away where they needed to ask him a few questions.

❖❖❖

The American Airlines security agent was actually a contractor from another company, an employee of ICTS, trained to uncover inconsistencies in people's stories, skilled at ferreting out lies, sensitive to subtle signs of nervousness. First impressions can be surprisingly accurate when you are observant, and the agent's first impression of the passenger was one of suspicion. It may have been this man's disheveled look, his standards of personal hygiene, his druggie gaze. Most likely it was a combination of things that gave rise to the hunch that not all was aboveboard. It was best to err on the side of caution, especially in view of recent history.

From the agent's and airline's point of view, the consequences of a *false alarm* — tagging an innocent passenger as a terrorist — were regrettable but tolerable as long as they didn't do it very often and they corrected their mistakes once they were made. A *correct rejection* was the term given to innocent passengers who were, as the phrase states, correctly rejected as being terrorists. Correct rejections were, by far, the most common classification of passengers passing through security.

Most airline travelers throughout the world only wanted to go about living their lives honestly and peacefully, not imposing their own values on others. Also, the airline didn't want to be perceived as making travel overly onerous for their good customers. The most important objectives were to accurately classify a security threat as a security threat, which they called a *hit*, and, of course, to never classify a terrorist as an innocent passenger, which was known as a *miss*. There had been 19 *misses* on September 11, 2001 in the United States, a fact that changed the course of modern history.

Bias was the term used to describe the shifting sensitivity to the signal or signals, the tendency of an observer to accept more *false alarms* or increase the probability of a *hit* at the expense of making more mistakes. Bias reflected the benefits and costs of all outcomes, but especially the consequences of missing a signal, which in this case was letting a terrorist slip through the cracks and past security. Bias shifted according to many factors, especially recent terrorist activity, the current perceived threat, and the observer's assessment of the danger to himself, his employer, his country, or even fellow passengers.

As for the *signal*, the stronger the better. The most important factor in detecting the signal was *sensitivity*, the capability of the system to actually detect what it was looking for, the resolution of the sensors, the ability to *discriminate* between an actual signal and background noise. In this case the noise was the ever-changing behavior of passengers, the things they could say, the objects and clothes they carried with them, their infinite variety of appearances and mannerisms and cultures that passed through the airport by the hundreds of thousands each and every day. No one ever pretended that detecting these people would be easy, especially with a system that may not be up to the task.

The agent already knew the answers to some of the

questions about to be asked, but they were asked anyway to see if the man's story matched the facts that he didn't know were known. "Where did you purchase your ticket?" "Where do you live?" "What is the purpose of your trip?" "Where is your luggage?" The ticket, the passenger replied, was purchased in Paris a few days before, a ticket to fly to Miami and then on to Antigua to visit his relatives. He had clothes and personal items at his relative's place. He didn't need to take any with him. He was Jamaican, he said, but a British citizen.

The security agent thumbed ever so slowly through the pages of his British passport, inspecting the photograph and examining each box for a departure and arrival stamp. The document was unusual in both appearance and content. The front half was new and had been issued only two weeks before in Brussels. An older but relatively new-looking passport was stapled to the back. It appeared to be missing a few pages and also contained a visa for a trip to Israel made in July. Israel seemed an odd place to travel for a man with no visible means of support, who occasionally worked as a dishwasher, a man with no permanent address and no appearance of having any purpose whatsoever to his life. There were no stamps in the passport showing a past arrival in Miami or, for that matter, Antigua.

The lack of luggage and the knowledge that the ticket had been purchased with cash were the most disturbing points. The agent's job was to detect passengers who might pose a security risk, a passenger having a motive and capability to hijack or destroy an aircraft with hundreds of people aboard. But as a contract employee of an airline, the agent had no authority to search the passenger or rummage through his possessions. The job had been done, though, and the scruffy man was asked politely to stay where he was for a few moments. The handset was lifted and a call placed. The words were spoken in a soft tone so as not to be overheard. An officer of the French Air and

Frontier Police would be available shortly to continue with the next phase of screening. In the parlance of *signal detection theory,* the passenger was a tentative *hit.*

❖❖❖

The ensuing discussions were held in private, away from the lines of people, and, with the exception of the French-accented English, they were nearly identical to the discussions held between the American Airlines security agent and the passenger. The outcome was considerably different, however.

"Why are you not carrying any luggage?" "Why are you traveling to Antigua?" "What are your travel plans once there?" The passenger had the same answers as before. His stories remained unchanged. His name did not show up on any terrorist watch list accessed by the French Air and Frontier police, nor was his name on a list of people with criminal records. He did not appear to be a threat, concluded the Air and Frontier Policeman, and, accordingly, there was no need to have him gone over by one of the six "explo" bomb-sniffing dogs on duty that day at Charles de Gaulle. And after an hour, and then more, the gaps in the conversation grew longer and the passport was examined and reexamined one last time.

Unlike his counterparts in the regular French National Police force, the Air and Frontier Policeman was forbidden by law from searching a passenger or examining his possessions. In fact, the passenger could not even be touched by him, and at half past 10 and without any additional evidence in support of the airline security agent's decision to hold him back, it was decided by the French Air and Frontier officer that the passenger should not be further detained. He was not a threat. Traveling without luggage was not illegal, paying cash for an airline ticket was not a crime, and although he pitied whoever had to sit next to him on the flight, you could not lock up a man for being malodorous.

From there Richard Reid, also known as Abdel Raheem and Tariq Raja, former British criminal who converted to radical Islam in a London prison, started the race to the gate, stepping prudently on the explosive-filled heels of his black suede high-top athletic shoes. Once again in the parlance of *signal detection theory*, he was now technically a *miss*, a terrorist who had been cleared through security and was about to board an airliner headed for America. But to his disappointment he soon learned that AA Flight 63 to Miami had already departed. He had missed his chance. At a nearby airline counter Reid was rebooked on AA Flight 63 for the next day, Saturday, December 22, and given a complimentary hotel room near Paris for the night as compensation for the morning's inconveniences, the fact that he had been, according to all this particular agent could tell, classified as a *false alarm*. His new seat assignment was in row 29, an ideal location, Reid knew, just at the back of the wing root and trailing edge, near an outside section of the fuselage and just over the aft end of the fuel tank.

❖❖❖

At 8:30 a.m. the next morning Richard Reid arrived again outside Terminal 2 at Charles de Gaulle International Airport. It was a drizzly December day. His damp clothes smelled even worse than the day before. After the previous morning's escapade, he had returned to the area near the Gare du Nord train station and Montmartre, settling in for the afternoon at an internet cafe within sight of the white domes of la Basilique du Sacré-Coeur, the Church of the Sacred Heart. He e-mailed his handler in Pakistan, who told him to persevere and proceed with his plans to reboard the flight on Saturday. From there he stopped at a fast-food restaurant and spent the night at an unknown location in the area. He did not check in to the hotel booked for him by American Airlines.

As on Friday, the CAPPS passenger screening system sent up a red flag on the agent's display when Reid checked in at the counter for AA Flight 63 to Miami on Saturday morning. Nothing had changed. He still had no luggage to speak of and was traveling on a ticket purchased with cash. But this time he sailed through the security check when he was singled out; the same security agent who interviewed him on Friday was on duty, and Reid was waved on. There was no point, it seemed, in having the passenger jump through all of the same hoops as the day before.

He went through passport control where the officer examined his photograph to make sure the picture matched the face and the name was the same as the name on the ticket. Given that Reid was departing France — not arriving — his passport was not scanned into the computerized system that might have shown where he had actually traveled in the recent past. The passport officer didn't know that Reid had acquired a new passport in July, telling British authorities in Amsterdam that his old one had been ruined when it went through the wash. Out with the wash went the obvious record of where he had traveled, the questions this might have raised, and his trips to Pakistan and Turkey. The Pakistan trip had an unrecorded crossing into Afghanistan and training at an al Qaeda camp in Khalden, near Kabul. His recent trip to Israel included a road excursion to the Gaza Strip and a jaunt across the border into Egypt. He also sent a letter from Iran to his mother. For a man who worked only occasionally as a dishwasher, he had certainly been getting around.

Reid had little concern about being detained at the mass security screen where the carry-on bags and coats were x-rayed and the passengers herded through the metal detectors like sheep passing through a gate. There was no metal in his shoes — just 10 ounces of triacetone triperoxide and pentaerythritol tetranitrate in the heel of each. The fuse that ran up the underside of each shoe tongue was nonmetallic as well. It was the same combination of powerful plastic explosives used to blow a deadly hole in the USS *Cole* the year before. Unlike the exploded al Qaeda bombs used elsewhere, Reid's bombs retained the personalized imprint of the palm of the bomb's maker and a strand of hair from someone other than Reid. Even if a metallic detonator had been placed in the plastic explosive in his heels instead of the chemical accelerator, it would not have set off the alarm; he knew the detector's sensor array was too high above his shoes to detect the presence of metal. Regardless, it no longer mattered.

❖❖❖

The Boeing 767 was now out over the ocean about seven miles up and equidistant from all runways on either side of the Atlantic, not quite halfway between Paris and Miami. Richard Reid was on the right side of the right aisle in seat 29H, one seat away from the window. The passenger seated next to him in 29J got up to go the lavatory. Reid shifted over to the now-empty seat to the right. He took off his right shoe and placed it on his lap, pulling back the tongue and exposing the fuse wire, slightly damp from his own perspiration and the wet Paris day. He then wedged the shoe between the right arm rest and the outside wall where it would blow a hole large enough to rip the plane in half. He pulled out his book of matches, tore one from the pack, closed the safety cover, ignited the sulfur tip on the scratch surface and held the flame under the end of the fuse running out the tongue of the shoe. Having flown exclusively on European and other carriers known for inconsistent enforcement of no-smoking policies on aircraft, Reid had no idea that striking a match on a U.S.-based carrier might create an immediate stir. From this perspective, an odorless butane lighter would have been a better choice.

Flight Attendant Hermis Moutardier, cleaning up the coach cabin after having served dinner, detected the whiff of smoke only seconds after Richard Reid struck his first match. She walked fast up the aisle and a passenger pointed to the man sitting in Seat 29J, next to the window. "Excuse me, you know that this is a nonsmoking flight!"

"Oh, I'm sorry, I'm sorry," replied Reid, who then stuck the lit match in his mouth to extinguish it.

Moutardier turned away and walked back 10 rows to the intercom where she called the cockpit and informed them that a passenger had lit a match, but that it was now out. She hung up the receiver and returned back up the aisle to check again. Incredibly, he had lit another match. His shoe was wedged against the wall and he was trying to set fire to the tongue with the flame. Then she saw the fuse. Moutardier grabbed for the shoe and Reid shoved her back hard. She came back at him fiercely, lunging for the shoe again. Reid, still seated and trying to hold the flame to the fuse to ignite it, slammed her to the aisle floor with his left arm. "Oh, my God!" she screamed at the top of her lungs. "Somebody help me! Water, contact solution, anything you have!"

Flight Attendant Cristiana Jones, drawn to row 29 by the smell of smoke, joined the fray, grabbing furiously for the shoe and the lit match in Reid's hand. She felt his teeth on her hand and the sharp pain of a crushing bite, and she screamed in pain. "Stop him, stop him! Someone stop him!" By this time a half-dozen bottles of water and plastic cups full of soft drinks had been handed up by nearby passengers, and Moutardier emptied a bottle on Reid and his lit match as he continued to bite Jones.

Eric Debry of Paris, sitting in row 30 behind Reid, awoke to the ruckus and screams for help and stood up and grabbed Reid's arms and shoulders. Two other men lunged for Reid's legs and, within a minute, Reid was being held down by a team of passengers while he screamed out in Arabic. The plane's relief pilot, taking a nap in the first row of first class, had by now been awakened and was orchestrating the permanent restraint of Reid. When his verbal request was not understood by the non-English speaking passengers nearby, he took off his own belt, held it up for everyone to see, and was passed twenty belts within a matter of seconds. More than a few headset cords were thrown in for good measure. Reid was securely lashed to the seat, and two large passengers stood guard over him for the remainder of the fight, one holding his mangy ponytail. Two physicians injected him with sedatives from the plane's medical kit.

Flight 63 was diverted to Boston, a shorter distance at that point in the flight, where FBI agents had some difficulty removing Reid from his make-do restraints. At his trial a year later Reid initially insisted that he had acted alone and out of his personal hatred toward the United States. He had learned how to make the bombs off the internet, he said. But by the end of the trial he had admitted to all eight counts against him and, with the fervor of a convert, declared publicly his loyalty to Osama bin Laden and membership in al Qaeda. On January 30, 2003 Reid was sentenced to life plus 110 years, with no chance of parole.

REFERENCES AND NOTES

Alleyne, R. (2003). Shoe bomber sentenced to 110 years. *Telegraph*, January 31.

Ashcroft, J. (2002). *News conference regarding Richard Reid [transcript]*. Washington, D.C.: U.S. Department of Justice, January 16.

Belluck, P and McNeil, D. G. (2001). A nation challenged: the suspect; officials remain uncertain on identity of suspect on jet. *The New York Times*, December 2001.

Canedy, D. (2001). A nation challenged: the travelers: a 'strange' traveler acted, and the passengers reacted. *The New York Times*, December 24.

Chaddock. G. R. (2001). Lessons of a shoe-bomb incident. *The Christian Science Monitor*, December 28.

Chamberlain, G. (2001). How did he slip through airline security? *The Scotsman*, December 24.

Cowell, A. (2001). A nation challenged: jailed Briton; the shadowy trail and shift to Islam of a bomb suspect. *The New York Times*, December 29.

Elliott, M. (2002). The shoe bomber's world. *Time*, February 16.

Exchange between Reid, judge follows life sentence (2003). *CNN.com www site*, December 6.

Explosives, wires found in sneakers (2001). *Milwaukee Journal Sentinel*, December 23.

French investigate security lapse (2001). *CNN.com www site*, December 25.

Joseph, F. (2002). Kwame James subdued British shoe bomber. *Newsday Trinidad & Tobago*, February 19.

Lavoie, D. (2001). Sneaker bomb suspect described. *Associated Press*, December 29.

McNeil, D. G. (2001). A nation challenged: the inquiry; French authorities wonder: how could it have happened? *The New York Times*, December 24, 2001.

McNeil, D. G. (2001). A nation challenged: airport security; bomb attempt has officials in France on defensive. *The New York Times*, December 27.

Moss, M. (2001). A nation challenged: technology; for airlines, yet another vulnerability. *The New York Times*, December 25.

Passenger: subduing shoe bomber changed life (2003). *CNN.com www site*, January 30.

Prada, P. (2002). Is the airport guard a European illusion? *The Wall Street Journal*, January 22, p. A14.

'Quite a scene' at sentencing (2003). *CNN.com www site*, January 30.

Reid: 'I am at war with your country' (2003). *CNN.com www site*, December 31.

Shoe bomb suspect 'did not act alone' (2002). *BBC News www site*, January 25.

'Shoe bomber' came within hairsbreadth of succeeding (2002).

Air Safety Week, August 12.

'Shoe bomber' first thought to be violating in-flight smoking ban (2002). *Air Safety Week*, November 18.

Sources: Reid is al Qaeda operative (2003). *CNN.com www site*, December 6.

Special report: coping with a crisis in the cabin: 'shoe bomber' flight triggers improved response procedures (2002). *Air Safety Week*, July 22.

Terrorist with shoe bomb exposes shortcomings in aviation security (2001). *Air Safety Week*, December 31.

Timeline: the shoe bomber case (2002). *CNN.com www site*, January 7.

United States District Court Criminal Complaint: United States of America v. Richard C. Reid (2001). Washington, D.C.: U.S. Department of Justice, December 22.

Who is Richard Reid? (2001). *BBC News www site*, December 28.

Young, W. (2003). *Ruling by U.S. District Court Judge William Young at sentencing of Richard Reid (transcript)*. Washington, D.C.: U.S. Department of Justice, January 31.

Zuckerman, L. (2001). A nation challenged: air travel; sigh of relief on Flight 63 and its effect. *The New York Times*, December 25.

OUT OF SYNCH

Olympic synchronized swimming judge Ana Maria Da Silveira Lobo, citizen of Brazil, native speaker of Portuguese, sat in her assigned seat in the box at the Olympic swim stadium in Barcelona awaiting the next contestant. The small input device to record her ratings rested on a ledge in front of her. The view below to the crystalline water in the pool was perfect. The contestants were the best the countries of the world had to offer. Next up was the favorite for the gold medal. You could feel the electricity in the air.

Unlike the classic 8-woman team event in which bronzed arms and shapely legs bloomed up from the water in synch to form intertwined patterns and woo the audience with impressive feats of aquatic athleticism in harmony with an emotional musical score, they were now on the solo event. The only synchronization was to be between the lone competitor in the pool and the music playing underwater and throughout the stadium. These were the compulsory routines, in which each competitor in the solo event had to perform four specific figures, as in figure skating, during a demanding two-minute program. The swimmer chose the music, but the compulsory maneuvers, or "prescribed elements" as they are called in the sport, were predefined. The "free routine" in which the swimmer could do whatever she wanted within the time allowed would be performed tomorrow. Each competitor had spent the better part of her life preparing for this day, and each competitor would leave Barcelona with the satisfaction of having made it to the Olympics. But only one would be judged — more or less — to be the best in the world and take home the gold medal.

❖❖❖

Ana Maria Da Silveira Lobo of Brazil was a member of the panel of judges ready to evaluate the *technical merit* of the next contestant's compulsory routine. Another panel judged *artistic merit* and would be rating the originality of the choreography, the fluidity of the transitions, the aesthetics of the patterns of legs and arms — the overall beauty of the presentation. For her part, Ana Maria had to focus on the slightly less-subjective and technical side of things. Her job was to verify that each of the prescribed elements was performed and evaluate the execution, synchronization, and difficulty of the whole routine. *Execution* encompassed the strokes with arms and hands and various thrusts with legs and feet, but also the precision of the patterns and twists and turns made by the athlete's whole body. The movements had to appear smooth and effortless, as if anyone could perform them, which, of course, was not at all the case.

The next competitor was Sylvie Fréchette of Montreal, Canada, native speaker of French, the pre-games favorite to win the gold. She was not only physically attractive but also exceptionally strong in both the upper and lower body. She would have to be high in the water in the all-deep pool during upright maneuvers, all the way to her waist on many occasions. When upside down, her legs had to be entirely above the surface so they could be seen by the judges and worldwide television audience. Synchronization with the music must be maintained both above and below the water surface, and the elements performed with her head underwater required her to hold her breath for as long as a minute while keeping the giant smile on her face for the underwater cameras.

The method used to measure athletic performance reflects the nature of the game being played, and this synchronized swimming competition was no exception. Determining which racing swimmer touches the wall first is a straightforward matter, especially with modern timing devices. The only thing that is required to determine the longest throw of the javelin is a very long tape measure, and for most team sports all the judges have to do is make sure everyone plays by the rules and then add up the number of times the ball goes into the goal. But things can get a bit tricky when *art* enters the picture, where the winning athlete is the one that *looks* the best or gives the performance that is *most pleasing to the eye.* Just how does one measure in a valid and reliable manner the thrust of a leg up into the air or a two-handed salute by a smiling young lady treading water? With a panel of judges, that's how, and a complex point system to help structure the process. Conceptually, each competitor started out with 10 points. If she was overly nervous and skipped one of her prescribed elements, a whopping two points would be deducted from the starting total of 10. The same was true for touching the bottom of the pool. Getting out of synch with the music could cost you a point, and a lesser offense like dropping the ever-present smile for a split second to get that gulp of air after a long underwater element might cost you a tenth or two if it happened to be seen by a judge. But these competitors were the best in the world and the differences between them would be a matter of hundredths and possibly thousandths of a point. *Restriction of range* is what the statisticians called it: for most competitors at this event the scores would be tightly bunched at the highest end of the scale.

❖❖❖

Sylvie Fréchette of Canada was now in the wings almost ready to start her performance with a ten-second ballet routine on the deck prior to jumping gracefully into the pool without so much as a splash. Somehow it was fitting that Sylvie Fréchette, the favorite to win the gold medal, was from Canada. The Canadians had, after all, invented the sport in the late 19th century. They called it water ballet and often performed it in turn-of-the-century indoor pools. For the participants it was a fun way to get some exercise and clean up at the same time; for the audience it was a pleasant and warm diversion on a cold Canadian day. But in the 20's the Canadians made the strategic blunder of showing off in the United States, and in 1934 a major water ballet exhibit was given at the Chicago World's Fair. The Americans knew a good thing when they saw it and embraced the sport as their own. Executives at the MGM movie studio in Hollywood created the aqua musical genre and made a star out of competitive swimming sensation Esther Williams. The United States Amateur Athletic Union recognized it as a sport, and water ballet was exhibited at the 1948 Olympics. Synchronized swimming made its full debut as a certified Olympic event at the 1984 Los Angeles Games, a stone's throw from Hollywood.

But as much as Sylvie Fréchette and Ana Maria Da Silveira Lobo and thousands of other competitors, judges, and spectators around the world loved their sport, it was not without its detractors. It was best left on the big silver screen, many believed, not elevated to the pinnacle of sport, the Olympic Games. Comedians couldn't pass up the chance to give mock performances on late-night television, especially during the "coming out" at the 1984 games in Los Angeles. And what was with the pounds of waterproof makeup, nose clips, glued-on grins, plastic hairdos, and high-cut sequined bathing suits? One sports columnist wrote that the *technical* and *artistic* dimensions of judging were best abbreviated, reduced to a basic form that said what the show was really all about: "T's and A's."

But comments like these were in very bad taste and ill-advised, especially in and around the synchronized swimming venue here in Barcelona in the summer of 1992. Sylvie Fréchette was ready to perform and the music began to play. She smiled to the crowd and to the judges, moved her arms and stood on her toes, and hopped into the water and began her difficult and physically demanding technical program at the Olympics.

❖❖❖

All had gone well and nearly two minutes later she was in the middle of her final prescribed element: an Albatross. Her torso was upside down and the top of her head pointed down to the bottom of the pool. Her waist was bent at 90 degrees, and her long legs were perfectly straight and parallel to, and just below, the surface of the water. Sylvie Fréchette sculled powerfully with her hands to stop herself from sinking. She raised her left leg straight up out of the water and pointed it to the sky, simultaneously bending her right leg at the knee and touching the upper part of the shin of her left leg with the toes of her right foot. The end of the routine was in sight. She straightened out her right leg and pointed it up to the sky right next to her left leg. Sylvie Fréchette was now fully upside down, yet, quite miraculously, half in and half out of the water. She eased off the pressure with her hands and slowly and gracefully slid down straight beneath the surface without a ripple. It was over! The routine had been completed on time. It had been a stellar performance. She knew it and the judges knew it.

A 9.7 out of a total of 10. This was the score judge Ana Maria Da Silveira Lobo of Brazil decided the performance deserved. All five technical judges had to enter their rating within a matter of seconds of each other. Each judge's small input terminal was connected to a central computer. Once Ana Maria entered her numbers they would be sent to the central computer and then back to the display on her hand-held terminal, verifying the receipt of her input. Long gone were the days when a judge had to pick up her stack of scoring cards, flip to the desired numbers and hold them up high at the same time as the other judges for all to see. The officials used to frantically write down the numbers, average them out, multiply a weighting factor for the technical and artistic elements, and come up with a final score while the anxious competitors and fans waited. Now a computer calculated the averages, applied the weightings, and generated the final score. It was much faster and less nerve-racking than the old manual method and, overall, it was less prone to error.

But there was still one aspect of the system where variability could creep in, where the vagaries of human behavior could play a role. Ana Maria's input device was much like a telephone keypad, with the numbers 0 through 9. There was a key for a perfect 10 and another for 1/2. There was no need to input a decimal point or press "Enter." These things were taken care of automatically. To enter a score of 9.7 she had only to enter 9 and 7. When all of the judges had keyed in their ratings the central computer would collect the values and start its calculations. What this required of Ana Maria was that she be exact and precise when she picked up the terminal and punched in her rating, just as she needed to be exact and precise when she used to flip to the appropriate numbers on the old manual scoring cards. You had only one chance and it had to be right.

She pressed the two buttons in succession. The values were registered by the central computer at the other end of the cable, and the numbers were sent back to the display on her small terminal. The display read 8.7. 8.7? 8.7 was not the score she had given Sylvie Fréchette! She had given her a 9.7. Ana Maria must have somehow mistakenly pressed the 8 key instead of the 9, just as one might do when dialing the wrong telephone number. Unfortunately, in this case she could not hang up and dial again.

In a panic she pressed the buttons on the keypad again, trying to erase her input and enter the correct score: 9.7. Nothing happened. The terminal would not respond. It had accepted her initial input as the score and there was no going back. Ana Maria knew immediately that she had to do something. She stood up and called for the assistant referee — Nakaka Saito, citizen of Japan, native speaker of Japanese — who came to her side. Their only common bonds were the love of synchronized swimming and limited knowledge of the English language. Nakaka Saito could not fully understand Ana Maria's Portuguese-accented English. She wanted a score of 9.7. 8.7 was incorrect and she wanted it changed. But the seconds ticked by and the spectators became aware that something was awry. Ana Maria Da Silveira Lobo's voice grew louder and more agitated. Finally, Nakako Saito recognized that she needed help keeping things afloat and she called for head referee Judith McGowan of the United States. Minutes passed. The audience grew restless. A decision had to be made. The score would not be changed, ruled the referees. 8.7 is what judge Ana Maria had entered into the keypad and 8.7 is what the contestant would receive. Scores from the other technical judges ranged from 9.2 to 9.6, and all scores were to be counted.

With Ana Maria Da Silveira Lobo's 8.7 and the various weightings and averages calculated by the computer, Sylvie Fréchette was awarded a score of 92.557 for the day, 251 thousandths behind her archrival Kristen Babb-Sprague of the United States. The Canadians filed a protest that day with FINA, swimming's governing body, but FINA backed the decision of the referees. The Canadian press cried foul, claiming that the head referee Judith McGowan, an American, had shown her true colors and bias in favor of the American competitor. Sylvie Fréchette still had the artistic performance tomorrow, but 251 thousandths was a large gap to close.

She did her very best the next day in the long free routine, scoring 5 perfect 10's from the large panel of judges, but it was not enough to make up the difference. Sylvie Fréchette graciously accepted her silver medal and listened politely to *The Star Spangled Banner* at the awards ceremony.

❖❖❖

Sixteen months later, after countless appeals by Canadian officials, the Olympic governing body ruled that Sylvie Fréchette, for her performance in the Barcelona Olympics, was to be awarded a duplicate of Kristen Babb-Sprague's gold medal. With her new title of Olympic Gold Medalist, Sylvie Fréchette moved on to greater heights in Las Vegas as the star of Cirque du Soleil's "O" Show, a revival of Hollywood's lavish water ballet performance of the big silver screen.

REFERENCES AND NOTES

Abrahamson, A. (2002). Precedent set for 2nd gold. *Los Angeles Times*, U7.

Blatchford, C. (1996). Swimmers water babes. *Toronto Sun*, August 3.

Farber, M. (1996). On the bright side: after tragedy and travesty in 1992, Sylvie Fréchette of Canada is thinking only good thoughts in Atlanta. *Sports Illustrated Olympic Daily www site*, July 30.

Fréchette, S. with Lacroix, L. [translated by Roth, K.] (1994). *Gold at last*. Toronto: Stoddart Publishing Co. Limited.

Sidney 2000: Synchronized swimming (1999). *Australian Broadcasting Corporation News www site.*

Silvie Fréchette: Against all odds (2000). *Life & Times [Canadian television show printed program preview]*, February 8.

Stubbs, D. (1996). Pooling her talent. *TIME International*, 148, 3.

DEATH ON CALL

The attacks on New York and Washington, the murder of thousands, and insertion of the Special Forces team in this godforsaken place were parts of a tale more bizarre than anything ever cooked up by Robert Ludlum or Ian Fleming. Who in their right mind a year before would have believed that a secret organization of 6,000 fanatics would set up base and take over a third world country, then carry out a devastating attack on the United States of America? No one would have believed it. No one.

But now it was all so real and so very believable to Captain Jason Amerine as they sat in the dirt on the hill once again another morning in the desert 18 miles north of Kandahar in southern Afghanistan. So much had transpired over the past month since they jumped out of the helicopters. It was good to have had a few hours of sleep for once and a few minutes to take the load off and let things settle down in your mind before picking everything back up again and making the final push.

The action had been what the team had trained for, yet difficult to believe it was true and that they might be close to the beginning of the end. The broad objective, knew Captain Jason Amerine, leader of the 11-soldier A-Team of Green Berets from the Special Forces Group of Fort Campbell, Kentucky, was the annihilation of al Qaeda and the defeat of their Taliban hosts. Amerine's immediate goal was finally wiping out the Taliban unit up in the orchard on the other side of the bridge over the

Arghandab River at the distant edge of the village at Showali Kowt spread out beyond at the foot of the little hill on which he now sat. The Special Forces unit and the band of friendly Afghans were on the brink of moving on to Kandahar, and it was time to deal with this situation once and for all and get on with the advance.

The break in the action came courtesy of the unit from headquarters flown in yesterday. Up until then Amerine's team and their small army of friendly Afghans were pretty much on their own, and they had done a very good job holding their own by any measure. But now, at 9:20 in the morning of December 5, 2001, the men of Amerine's team for once had the opportunity to observe rather than act. Another group and a superior officer had stepped in just as they were about to finally wrap things up here in Showali Kowt. Until last night they had not slept for days and there was no denying the intensity of the fights in and around the town over the past two days. In the end it was perhaps a good thing to have the relief and the additional personnel on site, or so it seemed at the moment.

❖❖❖

Not three months had passed since September 11 and only weeks since the Special Forces team had arrived, yet the overall American offensive had gone exceedingly well by any measure, beyond the most optimistic projections of the analysts at Central Command and not at all like the disaster forecast by the U.S. media. And who would have believed that the 200 or so Americans actually on the ground in Afghanistan had made it all possible. Within a matter of two months, the tribes of the Northern Alliance had come together with the help of the Americans and their high-tech weapons, al Qaeda and the Taliban were being pushed back into the corners of the country, and the northern two-thirds of Afghanistan were already on the road back to civility. A handful of U.S. allies had joined the cause, tons of equipment and munitions had been transported across the globe, and the aircraft were flying unthreatened high overhead. And now, in the first war of the 21st century, the satellites were up and working and the technology to put them to use in the making of war was in the hands of the best-equipped and most highly trained soldiers on earth.

This combination had been instrumental in the success of Jason Amerine's team and others like it on the ground in Afghanistan. It was a *system* of ships, thousands of support personnel, aircraft, satellites, guided bombs, and forward air controllers, a *system* that would prove to be even more lethal in the minutes ahead in the morning of December 5, here on a modest knoll of dirt and rocks overlooking Showali Kowt and the distant bridge and river and Taliban gathered in the orchard a kilometer away on the other side of town.

❖❖❖

This is not to say that the A-Team's work these past few weeks had been a cakewalk. The journey had been intense since the cold raw night they jumped out the doors of the helicopters into the choking cloud of dust and grit to meet up with none other than Hamid Karzai on November 14. Although the suave and articulate, multi-lingual Karzai was from the southern part of the country, he was the head of a band of Afghans from the Northern Alliance, and the CIA believed he had a good chance of becoming the next elected leader of Afghanistan. The Americans could further their own cause by helping Karzai, and this meant keeping him alive, something of ongoing concern considering the Taliban suicide squads sent out to kill him. Karzai had managed to dodge his assassins during the past month, although the same could not be said for other less-fortunate leaders of the Northern Alliance.

Karzai's friendly singsong British-Indian accent had sounded amusingly out of place under the roar of the retreating helicopters in the cold desert twilight that night weeks ago, but somehow honest and refreshing at the same time. Amerine felt that the brief initial meeting went very well. Still, when the helicopters had left and he looked around, things were not as he had planned. Instead of a few hundred well-armed fighters to complement the Americans' tactics, technology, and precision firepower, he and the 10 other men of the Special Forces unit now faced a bedraggled band of a dozen teenagers and men with a few hand-held weapons. And there were two donkeys that looked more like dogs, all led by a famous warrior who sounded like a British diplomat. There are times when you must have faith and go with what you are dealt, so after a few minutes on the ground they had loaded the two donkeys and begun a long hike under the starlight along a mountain trail to their next destination some miles away.

❖❖❖

The first test of men and weaponry was in the early hours of November 16th when the Green Berets and Karzai's Afghans set up to ambush an advancing column of 100 vehicles and 500 Taliban, mostly Arabs and non-Afghans, moving toward the town of Tarin Kowt from Kandahar through a mountain pass in a scene out of the old American West. Toyota pickups, flatbed trucks, and heavy artillery — not mules and a train of covered wagons — had entered the hemmed-in valley, and the Green Berets and Afghans were the guerilla fighters about to launch the surprise attack with the help of unseen aircraft high overhead. An offensive action such as this would have been suicidal only a few years before; some of the falling bombs would have wandered off target and landed on them as well as the enemy. They were also outnumbered by more than 10 to 1. But the new targeting technology had changed everything, or almost everything.

Each of Amerine's men was a highly trained specialist, including two combat control officers who were expert in operating their equipment to target enemy positions with either lasers or by sending GPS coordinates for a target to an aircraft overhead. With the laser marking system, they usually mounted their binocular-like device on a tripod, focused in on the target, illuminated it with the laser beam, and coordinated an airstrike with overhead aircraft; the falling bomb would then guide itself down to the illuminated target. The more modern GPS system was more complex and required extra pieces of equipment. First, there was the highly ruggedized hand-held GPS receiver, a PLGR (or Plugger, as the soldiers referred to it) made by Rockwell Collins Avionics of Cedar Rapids, Iowa. It was about the size of a standard brick. A stubby 6-inch antenna pivoted up off the right side and a short strap on the left side allowed the user to hold it tightly with the left hand and press the control buttons on the face with his thumb, keeping his right hand free. When turned on and after acquiring signals from overhead GPS satellites, the Plugger displayed its current location on the Earth, including latitude, longitude, and altitude — essentially the X, Y, and Z coordinates of a point in space relative to the center of the Earth.

When targeting for a GPS-controlled air strike, the Plugger was connected by a cable to a laser range finder, another heavy binocular-size device placed on a tripod. A target would be sighted and ranged momentarily with the laser beam. The distance to the target and the vertical and horizontal angle of the range finder were used by the system to calculate (1) the difference between the X, Y, and Z coordinates of the target and the X, Y, and Z coordinates of the Plugger, and (2) the GPS grid coordinates of the target, which would be displayed by the Plugger. This information was relayed by the combat control officer on the A-Team to the pilot overhead, who transferred it to the weapon about to be dropped from the aircraft. Once free of the aircraft, the ordnance "flew" itself down to the X, Y, and Z coordinates of the target that had been programmed into it. The so-called "smart bomb" was smart only in the sense that it had impressive sensing, memory, computational, and control capabilities. On the other hand, however, it was just as dumb as an old-fashioned free-falling iron bomb. Its target was based on coordinates in space and absolutely nothing else. It was the air controller's job to make sure that the coordinates for the strike were one and the same with the coordinates of the intended target.

Some components of the GPS targeting system had been received by Amerine's A-Team and other Special Forces units just about the time they had arrived in Afghanistan. Fortunately, there had been plenty of time to practice with the system in the days prior to actually putting it to use there above the valley approaching Tarin Kowt, learning its inputs and outputs, getting comfortable with the sequence of operations and the nuances of the device. As is the case with so many small hand-held devices, a lot of capability had been packed into the Plugger, and it took a good deal of time with the unit to understand the operation of the interface. It certainly did everything they required of it and more, but, really, who had need for a "man overboard" mode — to keep track of the position of a crewman who falls off a ship at sea — when you are in the middle of Afghanistan about to ambush a column of 500 Taliban?

The hint of daylight in the east signaled an end to the calm and the time to begin. Amerine instructed his combat air controller and weapons sergeant to transmit the positions of advancing enemy vehicles to the aircraft. Invisible laser beams from his team illuminated targets, attack aircraft dropped their ordnance from miles up, and the lead trucks in the convoy were suddenly blown to oblivion by precision-guided 2,000-pound bombs. There had not been time for Amerine to tell the friendly Afghans with them what was about to happen. When the explosions went off and the vehicles and enemy disappeared within balls of flame, the Afghans bolted from their positions, ran to their nearby pickup trucks, and made a beeline back to town. The Green Berets screamed for them to stop and come back. But with their only means of transportation deserting them, the Americans jumped on the trucks for the panicked ride back to Tarin Kowt. Once there, the team regrouped, returned to the ambush site in a handful of commandeered pickups, retargeted the advancing convoy, and called in devastating close air support with precision-guided munitions and low-level strafing. The fighting raged on for hours, and it had been a close call, but in the end over half the column of Taliban had been destroyed and the remaining enemy fled back to Kandahar in retreat. Three hundred Taliban had been killed, yet no Americans or friendly Afghans had been lost. Karzai would later call it the decisive battle of the war. Word of the victory spread fast and bolstered support for Karzai and his newfound American friends.

❖❖❖

In the days following November 16 the Amerine-Karzai team found its stride despite its small size. Karzai let the professional American solders take the lead when it came to orchestrating the battles, especially after he witnessed the initial pinpoint bombings of Taliban positions. The bombs, usually only one or two at a time, seemed to fall from out of nowhere, yet always obliterated the building or position where the Taliban were dug in. This was obviously not the same type of warfare fought by the hundreds of thousands of invading Soviets and their vulnerable tanks and low-flying aircraft two decades before. The contrast could not have been greater. The teams of Americans and their high-flying technology had accomplished in weeks what many believed could not be accomplished in years, if at all. On this battlefield each 11-member Special Forces team and their invisible friends in the sky were more powerful than a thousand Taliban with Kalisnokovs and RPGs.

Amerine willingly let Karzai take over all negotiations with local politicians, warlords, and even surrendering Taliban officers. This was Karzai's country, after all, and the Americans were not interested in staying so long as al Qaeda and the Taliban did not resurface from the caves. Karzai moved seamlessly between his native social environment and the brash, yet engaging, Americans. He was as comfortable negotiating the value of a man's donkey killed in a tribal skirmish as he was talking to United Nations diplomats on his satellite telephone about future Afghan foreign policy.

❖❖❖

Here and now, just after daybreak on December 5th, 2001 at the edge of Showali Kowt, Amerine looked back down onto the hundred or so mud brick buildings that made up the town that he had learned to know so well during the past 48 hours of exhausting battle. It is here where the Taliban had decided to make a stand after they had chased them for days. The American and Afghan allies had driven them out of town and over the bridge and across the river during fierce fighting on

December 3rd. The Taliban regrouped the night of the 3rd, however, storming back into the village and forcing Amerine and his men back up the little hill this side of town near where they started the day before. Fierce firefights followed the next day and one member of his team had been shot, but evacuated by helicopter and reportedly doing well. Once again they called in help from the guys up in the sky — this time a C-130 gunship and fighter-bombers who hit positions in the town and along the riverbank — and once again the surviving Taliban retreated back across the bridge.

❖❖❖

The superior officer from headquarters who arrived at Showali Kowt the day before had ordered them to take a breather and get some sleep early this morning. Not everyone was pleased about the headquarters unit dropping in and taking over, about being told to step aside and take a rest. They had handled every obstacle thrown at them and made tremendous progress. They had also become expert in their tasks, especially the operation of the new equipment for the air strikes. But it had been an intense two days and there had been no rest through any of it, and in the end Amerine felt better after the few hours of sleep before the sun came up. There also was much planning to do before moving on to Kandahar.

During the past few hours the new unit from headquarters had set up shop in a small white medical clinic building at the base of the hill below Amerine and his team. Karzai, some of his men, and a group of newly arrived tribal leaders had moved into a small school building next door to the clinic. Word had come from Germany that Karzai was to be named interim leader of Afghanistan. Taliban units were reportedly making surrender overtures in and around Kandahar despite the fact that the enemy unit on the other side of the river was showing no sign of leaving or giving up.

By 9:25 a.m. there were two to three dozen American soldiers and a number of CIA operatives scattered about the hillside and at the clinic building below. The new arrivals from headquarters had brought "care packages" for Amerine's unit: letters from loved ones and home-baked treats.

The Taliban had not left their position on the far bank of the river, and four enemy had been seen entering a cave about 1,000 meters away. And although they were not accurate, incoming rifle rounds occasionally peppered the American-Afghan position on the overlooking hilltop. The commanding officer from headquarters instructed his Air Force tactical air controller, who arrived in the country within the day and at the site only hours before, to bring in aircraft and hit the enemy position. With his tripod, laser range finder, and Plugger set up and an assistant at his side, the controller got his equipment powered up. The Plugger picked up the satellites overhead. The cave could be seen quite clearly through the powerful range-finding binoculars. He pressed the 12 keys on the face of the Plugger to negotiate the menus and selections, his only feedback coming from the little text display. He pulled up his own latitude and longitude in the form of grid coordinate numbers. His own elevation was also displayed. He had not had the benefit of the month of practice and real-world battle experience like the members of Amerine's A-Team. Things went slowly. Data traveled up the connecting cable to the Plugger GPS device. Finally, though, he pulled up the target on the display — essentially the grid coordinates for the degrees-minutes-seconds for the latitude and longitude — and then stepped through numerous sequences to display the target's elevation. A Navy F/A-18 fighter was approaching the area from a distance and waiting for this information to program its ordnance.

In the meantime, an Air Force B-52 bomber carrying 2,000-pound GPS-guided bombs had approached their location as well, and the Air Force controller on the hill realized that the B-52 would be on site minutes before the F/A-18. It meant, however, that he had to recalculate the target coordinates required by the system used by the Air Force bomber, a system based on *degree decimals*. They contacted the B-52 and arranged for the hit. The target was verified through the range finder once again as the B-52 approached. The buttons were pushed and the menus were scrolled and the selections made on the little 12-button keypad on the hand-held Plugger GPS. The controller had the target coordinates in the new units required by the B-52 when, of all things, the battery on the Plugger GPS went dead.

There was a small backup battery inside, but it was designed for backup power for core parts of the electrical system and memory, not the type of temporary data like the controller was dealing with at the moment. Quickly, the cap at the top right of the Plugger was unscrewed and removed, the dead battery slid out, and a fresh long cylindrical battery inserted and the cap screwed back on. The Plugger came back to life, its little screen displaying coordinates in degree decimals. The controller, unfortunately, was unaware that the GPS device's "active" coordinates had reverted to its *own* location — not the location of a target as seen through the range-finding binoculars to which it was still connected. And the display describing the position was, after all, just a bunch of numbers and interspaced periods, and the difference between his own location and a target location only 1,000 meters away, at least in terms of their appearance on a little display, was subtle indeed.

This had all taken a number of minutes, even more so than might be expected due to the battery swap and other things, and now the B-52 was in position. Without having the coordinates verified by his associate, the Air Force target controller relayed the target coordinates to the B-52 and called in the strike. The bomb bay doors opened high overhead, the ordnance fell free, the GPS antenna on the 2,000-pound bomb picked up the signals from orbiting satellites and calculated its location as it fell through space. The program came alive and the instructions were sent to the bomb's fins to steer it down to a point in three-dimensional space — a target which happened to be defined by an 11-meter-diameter circle of accuracy surrounding the Plugger GPS device attached to the range-finding binocular laser sitting on the tripod on the little hill overlooking Showali Kowt. The time was 9:30 a.m., the morning of December 5, 2001.

❖❖❖

A-Team's Master Sergeant Jefferson Donald "JD" Davis was passing out *Rice Krispies Treats* from a package sent by his wife and delivered by helicopter with the mail last night. Some members of A-Team were reading letters from home, and a number of Afghans and some of the Americans had taken the best vantage point on the hill to see the scheduled fireworks. "It really was kind of like Christmas," Amerine would later say. "...everybody was reading letters, sharing food with one another. Sharing it with the Afghans..." when the world came apart.

Amerine flew through the air for a sufficient length of time to know what had happened. When he finally landed in the dirt some distance away, he groped with his hands over his head, arms, body, and legs to see if all the parts were still there. His ear drums were blown out and his thigh opened up by shrapnel, but he was still in one piece, more or less. The same could not be said for Master Sergeant Jefferson Davis and Sergeant Daniel Pethithory of his Special Forces unit, his good friend Sergeant Cody Prosser from the other Special Forces team that had joined up with them, and about two dozen Afghans, all of whom were killed. The Air Force tactical control officer was not among the dead.

REFERENCES AND NOTES

Bender, B. (2001). Fighting terror: American casualties. *The Boston Globe*, December 6.

Bomb kills 3 U.S. soldiers, 5 Afghan fighters (2001). *CNN.com www site*, December 5.

Burns, R. (2001). 'Friendly fire' kills 3 U.S. troops. *Associated Press*, December 5.

Defense Department report, December 5: Afghanistan operations (2001). Washington, D.C.: Department of Defense, December 5, 2001.

Fog of war: facing friendly fire (2003). *CBSnews.com www site*, April 11.

Frontline: campaign against terror: interview with U.S. Army Captain Jason Amerine [transcript from television program] (2002). Boston: PBS/WGBH.

SOF Laser Marker (SOFLAM) AN/PEQ-1 (2002). *GlobalSecurity.Org www site.*

GPS award to Collins (1993). *Aviation Week & Space Technology*, April 5, 39.

Hendren, J. and Cooper, R. (2002). Fragile alliances in a hostile land. *Los Angeles Times*, May 5.

Hendren, J. and Reynolds, M. (2002). The untold war: the U.S. bomb that nearly killed Karzai. *Los Angeles Times*, March 27, A1.

Holt, D. (2003). Higher-tech bombs still not perfect: improvements cut friendly-fire risk. *Chicago Tribune*, March 20.

In Afghanistan with CPT Jason Amerine '93 (2002). *Association of Graduates (AOG), USMA, West Point www site*, March/April.

Jelinek, P. (2001). 'Friendly fire' kills 3 U.S. soldiers in Afghanistan. *Associated Press*, December 5.

Loeb, V. (2002). 'Friendly fire' deaths traced to dead battery. *The Washington Post*, March 24, A21.

McHugh, D. (2001). Hawaii native feels "numb" after Afghan bomb blast. *Honolulu Star-Bulletin*, December 9.

Naylor, S. D. (2001). Not victims, but heroes: members of his Special Forces team died fighting the Taliban. Now, telling the team's story, he wants to ensure them a proper place in posterity. *Army Times*, December 24.

News release no. 01-185, transcript of the telephonic press conference: Captain Jason Amerine (2001). Army Public Affairs 5th Special Forces Group (Airborne), Landstuhl, Germany, December 11.

News release: remarks by Captain Jason Amerine, 5th Special Forces Group (Airborne), commemoration service at Landstuhl Regional

Medical Center, December 12 (2001). U.S. Army Special Operations Command Public Affairs Office.

Orders for Rockwell Collins GPS receivers top 150,000 units: new order for 1,313 units received from U.S. Air Force [press release] (1999). Rockwell Collins.

PLGR+96 Precision Lightweight GPS Receiver, key features/user benefits (2002). Rockwell Collins.

Report: air controller called in friendly fire (2002). *Boston Herald*, March 27, 6.

Rumsfeld orders probe of "friendly-fire" bombing (2001). *USA Today*, December 6.

Scarborough, R. (2002). Karzai, A-team turned tide in November mission. *The Washington Times*, January 22, A1.

Shiel, A. P. and Smothers, A. D. (1998). Military GPS handheld display development: past, present, and future (paper no. 3363-53). *SPIE Proceedings*, 3363, 367-376.

Sources: 'friendly fire' airstrike likely human error. *CNN.com www site*, January 8.

Tragic turn — Pentagon: three U.S. servicemen killed in friendly fire incident (2001). *ABCnews.com www site*, December 5.

Woodward, R. (2002). Green Berets awarded for heroism in Afghanistan. *ArmyLINK News*, January.

PICTURE WINDOW

With his head aimed up (or was it down?) NASA astronaut Michael Foale floated weightlessly through the hatch into the *Destiny* laboratory module of the International Space Station (ISS), focused even more than yesterday, determined to find the source of their problem once and for all. They had spent almost every waking hour in the past 10 days trying to locate the air leak. Next to a fire or collision with another object, the greatest danger they faced while circling Earth at 17,000 miles per hour was the loss of their atmosphere into the vacuum of space inches away on the other side of the wall. Technical experts on the ground believed they detected signs of declining air pressure in the station, a situation which, if not addressed, could have dire consequences in time. Technically speaking, it was Sunday, Michael Foale's official day off, but there was no doubt in his mind that the only thing to do on this day was work the problem. The hunt for the leak had gone on for far too many days; now was the time to find the source of their troubles.

The first sign that air was escaping from the station came on January 2 when ground controllers in Houston concluded they were losing internal atmospheric pressure. Michael Foale and his lone fellow crewman, Alexander Kaleri of Russia, were told of the situation in calm tones during the next voice communication with the ground. The crew need not be alarmed, at least not yet, said the controllers, but, until they understood the problem and found a solution, other activities on the station had to be put on hold. What the ground controllers were seeing, Foale and Kaleri were told, wasn't an explosive fall in pressure, like what would happen if a valve or hatch blew out. To the contrary, this leak was slow, kind of like an air mattress on a camping trip that didn't seem as firm in the morning as it did the night before. It was a subtle problem but, nonetheless, one that would eventually jeopardize the crew and the station if they did not sort it out.

❖❖❖

ISS was designed to maintain a constant internal pressure of 14.7 pounds per square inch, the same pressure that exists at sea level down on Earth. The fall in air pressure had been both slow and small, so slow and small, in fact, that some Russian technicians in Moscow felt that what NASA had concluded to be a problem was, in fact, only the normal fluctuation in pressure expected of a large pressurized craft aloft in space. Everyone acknowledged that the station was never designed to be perfectly sealed like a glass sphere; its complex surface was comprised of welded joints and compressed gaskets and closed valves that would naturally lose a little air over time. The space station's air pressure, the argument went, rose and fell slightly all the time, especially in response to the alternating heating and cooling each time the 30-billion dollar football-field-size craft orbited the Earth 17 times each day flying alternately through the sun's bright rays and Earth's cold shadow.

But a careful check of the historical data beamed down to Houston proved otherwise. There was, in fact, a leak, and not only had they been losing air pressure since January 2, an analysis of old data showed that pressure had been dropping for at least 10 days before, for a total of three weeks. Instead of having the sea-level equivalent of 14.7 pounds per square inch, the International Space Station's pressure had fallen over these weeks to 14 pounds per square inch, the equivalent of an altitude of 2,500 feet above sea level. Perhaps the analogy to the leaky air mattress was not that far off the truth.

For more than a week, Foale and Kaleri had been checking the hundreds of possible sources — hatches, windows, ports, and vents. There was even the possibility that the space station had been punctured by a microscopic object as they orbited Earth, but this seemed unlikely in view of the lack of any other signs that they had had a collision, no matter how small. No, Foale and Kaleri both felt that the problem would ultimately be traced to a piece of equipment, perhaps a seal or a small part and hopefully, of course, not to one of the large main seals between the many large modules that made up ISS. Like tightening the tourniquet hard around a hopelessly mangled leg of an accident victim to save his life, they would have to close off one or more of the large hatches and abandon key modules in order to save the station if the leak was in a hatch or seal between two large modules. The two astronauts and everyone supporting them on the ground had worked somewhat nervously for days, hoping that the situation was not so acute that drastic measures like this would be necessary, but they knew it was a distinct possibility.

Upon receiving instructions from the ground, Foale and Kaleri had already closed off compartments at the remote ends of the station to begin the process of isolating the leak. All readouts showed that the pressure falloff had continued, meaning, unfortunately, that the leak was more centrally located and not within the remote areas they had sealed off first. In the meantime a bit of finger pointing was taking place back on Earth, where some in the American camp were suggesting that a Russian air-purification unit was the culprit: each time they switched it to a different operating mode there appeared to be a change in air pressure. As the undisputed masters of long-duration space flight, the Russians insisted that none of their equipment could possibly be responsible. Sometimes their designs may have appeared crude by NASA standards, but it was difficult to dispute the proven reliability of their basic components. But regardless of whom, if anyone, was to blame, they had made no progress solving the problem; NASA had come to a harsh decision, a decision for Foale and Kaleri to methodically close off modules and essentially shut down the entire International Space Station until they isolated and fixed the leak. This would all take place within the next few days. The grand space station habitat would become an uninhabitable 30-billion dollar piece of space debris within a few hundred hours if they were not successful.

❖❖❖

Unlike some of their American and Russian counterparts on the ground, Foale and Kaleri had gotten along splendidly throughout the duration of their extended flight in space. He was much too humble to say it himself, but Foale was widely regarded as an astronaut's astronaut: a man who possessed the requisite knowledge and skill to do a difficult and dangerous job and get along fabulously with every person he met and every situation he faced. His quick mastery of the Russian language and past space flights didn't hurt matters. Alexander Kaleri, known by friends as Sasha, was also highly likable and possessed the even temperament and global view that is so important in a long-duration space mission. Since the *Columbia* disaster and the grounding of the remaining space shuttles, Russian spacecraft had been called into service to transport crews and supplies to and from ISS. Crew size aboard the station had been reduced accordingly. All in all, there was perhaps no other pair of individuals in the astronaut and cosmonaut ranks who could have accomplished so much under such circumstances.

Now inside the American-made *Destiny* laboratory module, Foale floated to the main window, possibly the most optically perfect window ever made and most certainly the clearest window ever installed in a spacecraft. It was round and a little over two feet in diameter, and its surrounding circular framework was a seriously engineered structure encasing the layers of glass and seals. The window was the station's main viewing and photographic port for astronauts and cosmonauts to gaze at the Earth or out into space and take undistorted pictures with their cameras and high-powered lenses.

Each large window on ISS had to have multiple layers of glass for strength and safety. On any such window — especially a window to be used as a scientific viewing port — it was essential that the narrow space between the layers of glass be kept free of moisture and air and that there be absolutely nothing between the glass to impair visibility through the window. The growth of mold or mildew in the inaccessible space would render the window's special optical properties worthless. The Russian solution to this problem was to fill the inter-glass space with super dry nitrogen and seal it off, thereby keeping it dry and sterile, and the Russian-made parts of ISS made use of this design approach.

But the American-made *Destiny* module reflected a different philosophy, all due to the goal of creating the best window ever made. Instead of filling the inter-glass space with molecules of nitrogen gas, American designers filled the space with *nothing*. No, not air, and not an inert gas. *Nothing*. A limitless supply of *nothingness* was only inches away on the other side of the window. What better for filling the inter-glass space? Accordingly, next to the *Destiny* module's window at about the 10 o'clock position was a small port and valve, about the size of a small water faucet, linking the inside of the space station with the vacuum outside. A foot-long loop of strengthened metallic hose stuck out of the wall into the interior of the *Destiny* module, made a tight U-turn, and reattached to the hefty metal frame of the window a few inches away. This created a small but direct path between the vacuum of space and the inter-glass space via the U-shaped hose. The technical name for the loop of hose was a *vacuum jumper*. The result was the presence of *nothing* between the window's glass layers, a near-perfect vacuum where no moisture could accumulate and no organisms could grow to ruin the view through the world's most perfect window.

❖❖❖

Had it not been his day off, Michael Foale might have been floating leisurely in front of this very window on his own time, his camera held in his right hand, his left hand holding on to something — anything — to try to keep stationary and not float away when he snapped the shutter to capture a picture of the ever-changing view outside. Gazing out this window and snapping photos was the favorite pastime of most visitors to the International Space Station. Foale had already inspected the *Destiny* module on a previous day of their search, but Kaleri, who had good instincts, had a hunch that they should return for a second look. So now Foale was back again, going over the *Destiny* module for a second time.

And this time he held a special listening device, not a camera, in his right hand. It worked like a doctor's stethoscope and helped him hear the faint sounds of vibrations within objects via a headset. The objects were the hatches, walls, windows, frames, hoses, and machinery linked to the surfaces of the space station, surfaces that might conduct the faint sound of air escaping out a small hole or slit. With each contact of his probe to a solid object he hoped to hear the vibrations of a hiss or a whistle in his ears, the sound of air flowing out of their pressurized environment into the void. He listened for anything other than the constant spinning of small motors, fans, and parts that made an incessant low-level drone that might mask the whistle of escaping air out into space.

Then, quite unexpectedly, only days before the entire structure had to be sealed off and essentially shut down, just days before the 30-billion dollar enterprise would be declared inoperative, he heard it. Astronaut Michael Foale had the probe of the listening device touching the loop of silver hose that connected the frame of the picture window and its encased plates of glass to the small port leading to the vacuum on the other side of the wall. He touched the hose with his other hand and moved it gently. The pitch of the sound rose and fell in concert with each slight movement of his hand and the associated opening and closing of a small tear in the hose as the International Space Station's pressurized atmosphere escaped through it and out into space.

❖❖❖

In the days that followed, the Russians made public an expansive album of photographs spanning many many months. Each picture featured an astronaut or cosmonaut joyously absorbing the view of Earth while suspended weightlessly in front of the *Destiny* laboratory's picture window. In every case the right hand held a camera and the left clutched the silver-colored metallic hose connecting the vacuum of space to the panes of the best picture window ever made, the space-faring photographer holding steady as the shutter was snapped in the weightless environment of space. The hose was not designed to be a handle, but it was perfectly positioned for the purpose.

NASA acknowledged that a structural handhold and rack system designed to mount over the *Destiny* module window had never been put into orbit and installed in the International Space Station due to the grounding of the space shuttle fleet. Michael Foale and Alexander Kaleri installed a temporary replacement for the hose, and it was later replaced with a permanent part sent up to the International Space Station on a Russian spacecraft.

REFERENCES AND NOTES

No other leaks found on space station (2004). *Spacetoday.net www site*, January 19.

Oberg, J. (2004). Crew finds 'culprit' in space station leak. *MSNBC News Technology & Science www site*, January 11.

Oberg, J. (2004). Space station leak caused by crew, experts say. *MSNBC News Technology & Science www site*, January 16.

EVENT HORIZON

The young patient Michael Colombini, only 6 years of age, complained to his family of a headache. His appearance and words caused them to wonder if this was more than an ordinary headache from a cold or a touch of the flu. It was the kind of complaint that stirs unease in a parent, a hunch that everything is not as it should be, an anxiety about the health of one's child. These suspicions were confirmed days later on July 23rd when he fell to the ground for no apparent reason. Michael was in the hospital the very next day at nearby Phelps Hospital in New York. A cystic lesion on the right side of the brain showed up on the CT scan. Action was in order and he was transferred to nearby Westchester Medical Center for an MRI scan the morning of the 25th, with surgery that same afternoon.

The MRI of his head confirmed in greater detail what was suggested in the CT image. The neurosurgeons at Westchester Medical Center went about their important work soon thereafter. A right parietal frameless stereotactic-guided craniotomy and resection of the tumor was performed, cutting through Michael's skull in a precise location and removing the growth. It was a difficult and lengthy procedure, but one which had been completed many times before at the Center, a world-class medical facility serving a large populated area of New York. The initial results from pathology confirmed this was an astrocytoma, a tumor formed from astrocytes, a star-shaped neuroglial cell.

To the relief of his physicians and family, the outcome was very good. The tumor was benign; the prognosis was excellent. Aside from any setbacks or other events on the horizon, young Michael Colombini should recover fully with no loss of neurological function, his age working very much in his favor. He spent the night in the Pediatric Intensive Care Unit, monitored closely by specialists in Neurosurgery and Pediatric Neurology. The next evening on the 26th, with everything going well and Michael out of the woods after the surgery, he was transferred to the pediatric unit on the third floor for the next stage of his recovery. Michael was awake and alert, although in an understandable state of discomfort only a day after surgery on his brain. So far, though, it was American medicine at its best. Top-flight physicians had been able to diagnose and treat a condition thanks to state-of-the-art medical technology.

❖❖❖

At about a quarter after eleven the next morning, Michael was wheeled out of his room and down to the MRI facility where a follow-up scan after his successful surgery was to be made. Family members went along to comfort him, but Michael was still upset and crying despite the fact that he had had an MRI two days before and knew there was no pain involved in the procedure. He had been through a lot this past week and everyone understood his apprehension, but it was necessary for him to be calm and motionless while each image of his brain was made by the machine. The doctors thought it best to sedate him to calm him down and keep him still, so the anesthesiologist with him in the patient alcove of the MRI facility added the proper dose of drug to his IV. He quieted down soon thereafter, and the staff moved him onto an MRI-compatible stretcher so he could be taken into the room containing the massive MRI scanner.

Michael had been transferred onto the special MRI-compatible stretcher for a couple of reasons, the first being that once inside the MRI room he would have to be lifted over to the narrow MRI bed which would slide into the round center hole of the magnet just before the scanning began. The MRI bed was lower to the floor than most beds and the MRI-compatible stretcher was at the same relative height. Second, and most importantly, no ferrous metal could be brought into the MRI room. Any object containing iron — essentially anything that would be attracted to a magnet — would be drawn into the circular bore of the machine with great force if it got close enough. It was something akin to a black hole in the realm of astrophysics, those incomprehensibly dense collapsed stars in the distant universe that suck in everything, including particles of light that stray within their sphere of influence, the event horizon. The MRI scanner had its own "event horizon" and it extended out into the room many feet beyond the magnet's bore hole.

The MRI technologists and radiologists who worked with the scanner day in and day out were well versed in the importance of keeping ferromagnetic material away from the 10-ton magnet's bore. There were warning signs on the entrances to the doorways in the facility and only authorized personnel were allowed to enter. It was especially important to screen out patients who had any internal metal plates, clips, or pins that might be dislodged by the powerful magnet, and also necessary to keep things like steel wheelchairs and gurneys out of the room. Having the magnetic strips on one's credit cards erased or discovering that yet another wrist watch had quit working was one thing; a flying 50-lb wheelchair powered by a 1.5-tesla electromagnet was another.

The MRI suite was actually a series of rooms connected by

doorways. On the far end was the computer room housing the cabinets of powerful computers, data storage devices, and other expensive hardware used to analyze data and generate the images of the inside of the patient's body. The room that contained the scanner — the large doughnut-shaped magnet with a hole large enough to slide a person through — was referred to as the MRI room, and it was specially built and electronically shielded with sheets of metal in the walls. Outside the door of the MRI room was the small patient alcove and the console room where two MRI technologists sat in front of keyboards and visual displays. A window in the wall between the MRI room and the console room gave the MRI technologists a direct line of sight through to the bore of the magnet and the patient during the procedure. The expensive machine was typically booked through all day with appointments every 45 minutes or so, and the technologists usually focused on the equipment, the procedures, and taking the scans rather than on the medical condition of the patient; that was the job of the attending nurse or physician, which in this case was the anesthesiologist administering the sedative to Michael.

❖❖❖

Now on the MRI-compatible stretcher, Michael was taken into the MRI room and lifted onto the narrow bed that would slide into the magnet's bore. The taurus-shaped magnet at the heart of the multimillion dollar machine in the MRI room was about as high as a tall man and encased inside a modern cream-colored shell. The powerful magnetic field in and around the bore would cause individual atoms, especially hydrogen atoms contained in fat and water, to resonate. Harmless radio waves would then be introduced into the field, and the resonating atoms would give off radio energy that was measured by receivers also contained within the structure. Multiple images would be taken, each requiring that Michael be perfectly still for a half-dozen minutes. The peculiar but normal knocking sound as field gradients turned on and off within the cryogenically cooled magnet should not disturb him in his sedated state. He would probably remain quite motionless, and the resulting images would be especially useful for looking past bone, into the soft tissue within Michael's head, to see the aftereffects of the surgery. The processed images would appear on the technologists' screens and then be printed on film for later viewing by the radiologist and neurosurgeons.

Michael, like many patients in a postoperative or sedated state, needed to have supplementary oxygen piped to him while the scans were under way. Accordingly, two large oxygen cylinders, each with a pressure gauge for measuring psi, were strapped to the wall inside the computer room, some distance away from the magnet. A concealed tube ran from there to the wall of the MRI room where a flowmeter was located to regulate how much oxygen was moving through to the patient. A flexible tube from the valve at the wall delivered the oxygen over to the narrow bed in the scanner. The anesthesiologist opened the valve to start the flow of oxygen at 4 liters per minute.

Sensors were attached to Michael to measure the oxygen saturation in his blood, his heart rate, and the carbon dioxide in his breath. With his head held in place by a supporting jig, he was slid carefully into the tunnel of the magnet and positioned in the necessary spot. The technologist left the room, closed the heavy door, and went to the workstation in the console room. The anesthesiologist stayed in the MRI room, sitting by Michael's side to monitor his vital signs throughout the scan. The MRI was harmless, so there was no danger to him in staying nearby.

❖❖❖

The imaging was to begin momentarily. Michael was lying still, his wrapped head inside the bore and the rest of his small body extending out on the narrow bed. The patient and the machine all seemed in order, all the procedures had been followed, and yet something did not seem quite right within the room. Perhaps it was the changing color of his skin or the unusual silence. The anesthesiologist looked down at the oxygen flowmeter. It read 0 liters/minute. He opened the valve further and the needle did not move. The tank must be empty! They were supposed to be switched out when down to 500 psi. The pressure must have been very low when they first started getting ready, and now some minutes later the tank was empty. They had to switch to the other tank, quickly!

He had to stay with the patient, there was no question about that. Someone else had to immediately switch the tanks back in the computer room. There was not a microphone in the MRI room with which to talk to the technologist in the console room, so the anesthesiologist jumped from his stool and stepped to the window and wrapped it repeatedly with the knuckles of his hand, getting the attention of the technologist on the other side. The door to the MRI room opened. "No oxygen was flowing to the patient from the tank in the computer room," the technician was told in a raised voice by the anesthesiologist. There was no oxygen! Without hesitation the technologist turned and went back through the opened door, moving quickly across the console room to the computer room. The second technologist in the console room also went into the computer room to help switch the tanks. The door to the MRI room remained open, as did the nearby door between the console room and the hallway outside.

The anesthesiologist grew increasingly concerned with each

passing second. He yelled out for the technicians to hurry up. At that same moment a nurse who had been in the MRI suite was walking by the MRI room. The door was wide open, and she heard the anesthesiologist's urgent calls for oxygen.

There, just a few feet away in the patient alcove across the hall from the door of the MRI room where she stood, was a hand cart with two size E oxygen cylinders, each about the size of a small fire extinguisher. She bent down and looked at the pressure gauges. The first read empty, the second was 3/4 full. She lifted the 20-lb second tank out of the hand cart.

The anesthesiologist, still in the MRI room with his patient, approached the doorway 15 feet from the magnet's bore. The urgency of the situation was clear to all involved. This could be a life-and-death situation and quick action was needed. The warning signs on the walls and general knowledge that they each possessed was not enough to overcome the immediacy of the need and their desire to act quickly to protect the young patient's health. For all practical purposes, the greater danger was entirely invisible. The forces at hand did not glow like flames of fire or give off any sound.

Like two well-trained soldiers passing a live artillery shell from one to the other, from the ammunition rack to the breach of the cannon, the 20-lb tank of oxygen left one set of hands and was grasped carefully and expeditiously by the hands of the second on the other side of the doorway. The anesthesiologist turned.

At a point in space, somewhere along the surface of the invisible three-dimensional sphere that extended from three to six feet out from the bore of the immensely powerful electromagnet, the 20-lb steel oxygen tank passed through the "event horizon" of the MRI's magnetic field. There was no turning back. The 20-lb downward gravitational force acting upon the tank was exceeded by the lateral suction of the magnet.

Milliseconds thereafter the pull on the tank by the magnet was hundreds and hundreds of pounds, far exceeding the strength of any human, and in far less time than it would be possible to react the tank shot out of his hands as if blown out of the barrel of a cannon. He frantically swiped and grabbed, but by then there was only thin air. The oxygen tank shot straight for the center of the round MRI magnet bore where it slammed into Michael Colombini's skull with lethal impact. Two days later, after initiating brain-death protocol, Michael was pronounced dead at 5:40 p.m.

❖❖❖

With candor highly uncharacteristic of medical institutions, Westchester Medical Center announced the MRI accident the next day, and undertook a comprehensive and open investigation into its causes. "The Medical Center assumes full responsibility for the accident, " said President and CEO Edward Stolzenberg. "Our sorrow is immeasurable and our prayers and thoughts are with the child's family. The Medical Center will do anything it can to ease the family's grief."

The report on the investigation was made available to the public one month later. Blame was placed on hospital systems rather than individuals. "This is more a failure of hospital systems than a failure of people," stated Stolzenberg. "I cannot in good conscience blame any of the healthcare workers who were each trying to help the patient." "If healthcare institutions want the public's trust, they must earn it by admitting mistakes and learning from them. If we can save one other life with our admission, we have to try. Our hearts will always be with the Colombini family."

Among other things, it was determined during the investigation that there was a need to enhance the communication systems between the MRI room and the console room, revamp training and education of all personnel having any contact with the MRI facility, restrict access to the MRI facility, expand the restricted magnetic field area to include the MRI patient alcove and immediate surroundings, create physical barriers serving as visual reminders for personnel, and prohibit the use of ferromagnetic oxygen tanks anywhere within the MRI suite, not just the MRI room. If oxygen was to be used by a patient undergoing a scan, a full aluminum tank would be brought into the MRI room and monitored at regular intervals. And in an interesting twist it also came to light during the investigation that an oxygen tank had been sucked into the MRI bore four years before, although no patient was present at the time.

A somewhat cavalier attitude about MRI safety had developed over time with hospital staff — especially personnel not involved in the day-to-day operation of the MRI facility — unaware of the potential dangers of the technology. On the day of the accident a cart with ferromagnetic materials sat in the MRI patient alcove and the fire extinguishers on the wall of the MRI room were made of iron-containing steel. Fortunately, there had never been a fire requiring their use.

Taking the case a step further, Westchester Medical Center had Dr. Emanuel Kanal, a world-renowned expert in MRI safety, conduct an independent review of the accident, the internal investigation, and ensuing countermeasures. The Center also asked the American College of Radiology to convene a meeting of top engineering, scientific, and medical personnel in the industry to develop policies and guidelines for safe MRI practice. Westchester Medical Center's response to the tragic event greatly enhanced the cause of MRI safety throughout the nation.

REFERENCES AND NOTES

An horrific MRI accident occurred at our hospital and has resulted in the death of a young child [news release] (2001). Westchester Medical Center, July 30.

Child's death from MR injury closely follows report on MR safety (2001). American Society of Radiologic Technologists, August 7.

Fitzgerald, J. (2001). MRI accidents may be on the rise. *Associated Press*, July 31.

Gosbee, J. and DeRosier, J. (2001). *MR hazard summary*. VA National Center for Patient Safety, August.

Hospital releases report on MR fatality, asks ACR to update MR safety guidelines (2001). American Society of Radiologic Technologists, August 31.

MRI accident kills 6-year-old boy (2001). *Santa Barbara News-Press*, July 31, A4.

MRI expert gives high marks to Westchester Med Center's new safety measures: explains physics of July accident [news release] (2001). Westchester Medical Center, August 31.

Westchester Medical Center releases internal MRI investigation report: Medical Center calls for creation of nationwide patient safety standards [news release] (2001). Westchester Medical Center, August 21.

WCHCC incident review of July 27 accident (2001). Westchester Medical Center, August 21.

FREEWAY DRIVER

Traffic on the 110 Harbor Freeway was heavy, but moving fast at 70 miles an hour pointed north to downtown Los Angeles. To the left was the old Sports Arena and Coliseum and the island campus of USC out in the sea of asphalt and chain link. Then around him and above him was the interlaced interchange of the 110 and the 10 Santa Monica Freeway. Off to the right in a minute was the Convention Center and Staples Center, home of the NBA Champion Lakers. The elevated freeway gradually lost its height and dipped under the web of downtown streets. First came Seventh Street, followed fast by Wilshire and Sixth Street and Fourth Street and Third Street, where the ponderous overhead green aluminum sign pointed down to the left three lanes for the 110 Pasadena Freeway and the center two lanes for the northbound 101 Hollywood Freeway and the right two lanes for the southbound ramps to the 5 Golden State Freeway or the eastbound direction of the 10 Santa Monica Freeway, if that is where you needed to go. City Hall and the Music Center, site of the Oscars, slipped by on the right just as he dipped under the 101 and found himself on the Pasadena Freeway, but still on the 110. Then in and out of four short tunnels at 70 miles an hour with cars on all sides, ahead and behind. And as if out of nowhere, what he was looking for and the lane in which he now needed to be at this very moment in time was off to the left — *all the way over* to the left; the exit to the northbound 5, the Golden State Freeway, marked by a separate sign located way too close to the exit in both distance and time. Aaaaaah! Southern California. Home of the freeway. And so much for the exit to the northbound 5.

For Richard Ankrom — avant-garde artist, part-time sign maker, freeway driver, and resident of Los Angeles but native of Seattle way up north over 1,000 miles away near the distant end of the 5 Freeway — this was not the first time the exit to the northbound 5 off the northbound 110 Pasadena Freeway just past downtown Los Angeles beyond the underpasses and tunnels of concrete and steel had snuck up on him like a rattlesnake in the dry grass. He had spent countless hours plying L.A.'s freeway trails and knew the demands that were placed on one's memory and skill when the traffic moved along at its top rate instead of a crawl. There wasn't time to look at a map or read the signs. You had to know everything by heart and have a map in your head or memorize your plan before setting out. Yet this particular transition always caught him ill-prepared and in the wrong place, even though he had driven it many times. There just was something about it, that the sign was so small and off to the side, and that you came upon it without warning at the last possible second after those other overhead signs diverted your attention away from what you were looking for in the first place. Yes, it was true that some of this was to be expected while driving the freeways, but this had to be about the most user-unfriendly signage around. There had to be a better way.

He could write a letter or call Caltrans, California's monolithic Department of Transportation, responsible for the design, placement, and maintenance of freeway signs in this state of 36 million people with the sixth largest economy in the world, right behind Germany and France and one slot ahead of the U.K. Right. Dream on. Even after countless calls and endless meetings with administrators and engineers and announcements for public comment and hearings, nothing could possibly ever happen. Why would they bother listening to him? What did he know about freeway traffic theory, speed-density relationships, kinematic waves and traffic flow, or the Manual on Uniform Traffic Control Devices for Streets and Highways? Nothing, of course. Zero. Nada.

Yet the problem had to be so obvious to someone driving through the interchange. Anyone would be perceptive enough to see the random smattering of brake lights and vehicles jolting across multiple lanes of the six-lane freeway when that first warning for the northbound 5 appeared on a small pole sign, at a point when it was essentially too late to do anything about it if you were not already over to the left. You didn't have to be a traffic engineer to see what was going on, and complaining to Caltrans was not the answer, since the situation had been like this for years and they had never done anything about it. Perhaps there was a more direct solution, one that relied on his own talents.

And so it began for Richard Ankrom, freeway driver, avant-garde artist, and sign painter *extraordinaire*. He embarked on a mission to rectify a wrong, a project to ease life for the masses, a selfless act for humanity, a calling — when you got right down to it — to create and display *a work of art* that would be seen by 150,000 people each and every day of the year, more than the hordes of humanity standing in line to get into the Louvre and as many admirers in a single day as shuffle past the *Mona Lisa* throughout an entire month. A work of art that could redefine *work of art*, a painting that would become the most widely viewed work of art in all of history!

❖❖❖

The first thing to get a handle on was the objective — not in terms of art — but in terms of the freeway driver. What would it take to get drivers headed for the northbound 5 exit in the right place at the right time? To begin with, a driver had to know that there was an exit up ahead, and, secondly, you had to know which lane you had to be in when the time came. Most importantly, you needed to know all of this some distance ahead of the exit so you had time to move over to where you needed to be, and this, of course, was the fundamental problem with the current setup. There was such a short distance between the small sign for the northbound 5 and the actual exit. If the traffic was moving unobstructed and you were not already in one of the left lanes, it could be very difficult — if not impossible — to change lanes over to the left in the available time and distance. It was also a very complex interchange, with forks and exits to the northbound 110, the north and southbound 101, the southbound 5 and the east 10. What was needed was another sign for the northbound 5 exit a mile or two further upstream, giving you time to get over to the far left lane.

❖❖❖

So Richard Ankrom drove the northbound 110 again, and then again. The most logical spot for a sign for the exit to the northbound 5 was on the large dark green backboard on the sign gantry that ran all the way from left to right above the six lanes just after the 110 ducked under the Fourth Street overpass two miles before the exit to the northbound 5. The leftmost portion of the sign over the left three lanes read "110 Pasadena," with reflective upper and lowercase letters in white on the green background. Three white downward-pointing arrows near the bottom edge of the sign pointed out the three lanes — lanes 1, 2, and 3 — from left to right. There were also two smaller "NO TRUCKS" signs, with dark characters on a white background, mounted between the arrows for lanes 1, 2, and 3.

The sign above lanes 4 and 5 read "101 NORTH, Hollywood." There were also two downward-pointing arrows to the two lanes and a smaller "TRUCK RTE" sign as well. Over the far right lanes another segment of the sign read "5 SOUTH, 10 EAST, TRUCK RTE," with arrows pointing downward to the appropriate lanes.

In addition to being upstream of the actual exit to the northbound 5, the good thing about this location near Third Street was that the leftmost segment — the segment providing directions to the three lanes headed to "110 Pasadena" — was visually clean and not cluttered with numerous words or numbered highway shields. In fact, it was quite spacious, and the left side of the sign, the side over the leftmost lane, had plenty of room for a — *work of art*. It might not be the *Mona Lisa*, but it would be a work of art none the less.

❖❖❖

If it were going to be done at all it had to be done right, so much so that it would appear to everyone that this was the way things had always been, that nothing had been changed, but somehow, someway, the interchange to the northbound 5 Golden State from the northbound 110 Pasadena Freeway worked a little better than it used to. What was required was actually quite simple. First, he needed a so-called "shield" sign, one having the distinctive outline designating the 5 Freeway as part of the federal system of interstate highways. The horizontal segment across the top of the shield must have a background in red with the word "INTERSTATE" in relatively small white characters. The main body of the shield sign would have to be in dark blue with a narrow white outline to set it off from the large green billboard-size sign on which it would be mounted, and the

"5" would have to appear in reflective white, right in the center of the blue background.

A driver would also have to know that this was the direction for the northbound 5, so a second sign would be required, one that said "NORTH." The new "5" shield sign could be mounted over the far left lane, about halfway between the bottom and top of the large green background, and the "NORTH" element placed just above it, as on the thousands upon thousands of similar signs across the interstate highway system. And that was it, that was all that was needed: a "5" shield and a "NORTH" sign. There was plenty of room to add the two new elements; the additional information would not conflict in any way with negotiating any other part of the interchange, and a new "NORTH 5" addition over the far left lane would create a logical counterpart to the "SOUTH 5" labels that were over the far right lanes.

❖❖❖

In the following weeks and months Ankrom photographed and measured signs throughout the Los Angeles freeway system, getting an understanding of the rules that were generally applied, the conventions of placement of the various elements. And, surprisingly, he found that all of the detailed information for the design of these signs was readily at hand courtesy of the Federal Highway Administration's internet site. Chapter 2E of the Manual on Uniform Traffic Control Devices for Streets and Highways contained 82 pages on everything you ever wanted to know about freeway sign design requirements but were afraid to ask, including shape and size, color and reflectivity, letter aspect ratio and stroke width, and even the thickness and type of aluminum sheeting for the sign itself. Perhaps this shouldn't have been so surprising, on second thought. These weren't the government's top-secret specifications for the latest experimental cruise missile, after all, only logical and well-done design guidelines that could be accessed and used by any company wishing to get into the sign-making business or any traffic engineer needing information to replace a sign or have a new one made. It was in the interest of everyone to have clear and legible signage, and it made sense that this information was entirely available to the public.

Ankrom took his time, scouting materials suppliers, finding the right kinds of paint, sketching out his signs and placement alternatives, even projecting his photographs on a wall to get a sense of just how surprisingly large his new signs would have to be. Once ready, he acquired a big sheet of 0.080 mm 5053 aluminum and cut out his shapes. A coat of zinc chromate primer came first, followed by meticulous masking and application of Pantone red 199-200, blue 293, and green 340-341 with a spray gun. The white reflective letter "5" for the interstate highway shield and the "NORTH" letters required large button reflectors according to the federal specifications. Avoiding normal supply channels so as to avoid raising suspicions, he located a stockpile up in Tacoma, Washington, telling the inquisitive order clerk he needed them for a movie. The finished signs looked good — perhaps too good — so he applied a light layer of gray paint to give them an aged patina and tone down the colors so they would not stand out from the existing signs on Gantry 23100 over the northbound 110 Freeway.

❖❖❖

Two years later the moment was at hand, two years of research, craftsmanship, cajoling and planning, two years of attending to every detail he could think of so the whole thing would go off without a hitch. His project was no longer a secret, although Ankrom's small cadre of fellow artists and friends, who knew what he was up to and helping out where they could, were sworn to secrecy.

With only a few days left until D-day, August 5, 2001, the final tasks were checked off one by one. Ankrom's shoulder-length blond hair would be a dead giveaway to any passing California Highway Patrol officer or Caltrans worker, so a friend give him a haircut, of sorts. They scouted the overhead sign gantry near the Third Street bridge and developed a plan for getting past the razor wire placed there to discourage graffiti artists. He decided how he would use a ladder to get up onto the shallow work platform in front of the sign spanning the freeway. The installation process had to be quick but look normal to the thousands of motorists who would see him, and he thought about how he would carry the signs across the work platform all the way from the right side over to the left above the vehicles speeding by below. It was apparent that he would need a stepladder for placing the "NORTH" sign since it was going up near the top about three feet above his head once he was up on the platform. He prepared a temporary template to hang onto the green background to make certain that his "NORTH" sign and "I-5" shield were straight and appropriately spaced when he drilled the screws into the green signboard.

On Saturday morning August 5, 2001 Richard Ankrom arrived at the Third Street bridge above the northbound 110 Freeway in Los Angeles and parked his pickup truck. His friends were in position to secretly photograph and videotape the event. He donned an orange traffic safety vest and white hard hat and placed two orange traffic cones next to the truck. His pocket held a bogus work order for the freeway sign installation. A logo had been stenciled to the door of the pickup. To any passerby it looked like most every other construction company banner, but a closer look might have raised suspicions. It read: "Aesthetic De Construction."

The walkie-talkies were on. No suspicions had been raised. So far, so good. Word came from one of his hidden accomplices: "Move in, Rubber Ducky."

Ankrom worked quickly but deliberately like any blue-collar construction company or state employee who wanted to get the job done and spend as little time as possible perched 20 feet above the dangerous freeway traffic. Much of the gear was already pre-positioned in the bushes next to the freeway. He leaned a tall ladder up against the gantry. Ankrom grabbed what he could carry up, negotiated the razor wire, and stepped onto the elevated work platform. Cars and trucks and buses and motorcycles roared along at 70 miles per hour just beneath his feet and the see-through metal grid platform. He walked over the freeway, over the lanes that would take drivers to the southbound 5 or the eastbound 10, then over the lanes to the northbound 101, and then to the three left lanes headed to the 110 Pasadena Freeway, and finally he was there, above the far left lane that would eventually take you to the northbound 5. He bent down and carefully raised a folding safety rail at the edge of the work platform, then attached his temporary mounting template to the green backboard and raised his stepladder. The "NORTH" sign came first, which he held in place and screwed into position with a cordless electric drill. Then the all-important "I-5" shield was screwed into place. He picked up all of his tools and materials, carefully lowered the safety rail back into its stowed position, picked up the stepladder, and worked his way carefully across the platform above the speeding vehicles, over to the right side and then down to the ground. It was then back up the embankment and a quick walk around the corner to the Third Street bridge. He put his tools back into the bed of the pickup, threw in the orange traffic cones, got in the truck and drove away.

❖❖❖

In addition to his normal job painting signs in and around Los Angeles, Ankrom spent the next nine months working on his other art projects and editing the film taken by his friends. By this time over 40 million pairs of eyes had passed under the big green sign on the northbound 110 Pasadena Freeway with its new "I-5" shield and "NORTH" sign over the far left lane. The traffic did seem to flow better leading to the transition of the northbound 5 Golden State Freeway two miles downstream, and the new signs certainly didn't appear to have caused any problems. The *thing* was, though, no one seemed to have noticed — at least on a conscious, public level. He had never been contacted by the police about his illegal sign installation stunt back on August 5, Caltrans had not touched his artful additions, and there had been absolutely nothing in the newspapers, television, or radio about the changes to the signage on the northbound 110. But this was, after all, the true functional point of the project, to make a public work of art that benefited the public, "...to manufacture and install these missing guide signs to ease the confusion and traffic congestion at this section of the 110 Freeway," said Richard Ankrom. Fame was, if it should come, only a by-product of his effort and should not be of concern.

But fame came, of course, just days before the scheduled showing of *Guerilla Public Service* at a small film festival in Los Angeles. A friend had spilled the beans to the press. First came the article in *LA Weekly Magazine*, followed by the big piece in the *Los Angeles Times*, then the bits on ABC, CBS, CNN, and in *USA Today*. Caltrans could only say that "Mr. Ankrom did a fantastic job." The sign was made and installed to perfection, and it would be left in place. No charges would be filed against Richard Ankrom. Yes, this was Southern California. Home of the Freeway — and so much more.

As for the artistic merits of this most-viewed painting in the world, there was still some controversy, however. In one of many 'letters to the editor' of the *Los Angeles Times* regarding Ankrom's antics, one reader from Covina wrote: "Maybe the clandestine art project eluded official notice for months because it's not art, it's just another freeway sign."

Regardless, it was difficult to argue with the result.

REFERENCES AND NOTES

Ankrom.org www site (2002).

Ankrom, R. (2002). *Guerilla public service [art video]*. Produced by Richard Ankrom.

Cullum, P. (2002). Guerrilla public service: the man who would be Caltrans. *LA Weekly.com www site.*

Chapter 2E. Guide Signs: Freeways and Expressways. Manual on Uniform Traffic Control Devices for Streets and Highways (2000). Washington, D.C.: US Department of Transportation, Federal Highway Administration.

Martin. J. (2002). In artist's freeway prank, form followed function. *Los Angeles Times*, May 9.

Pawluk, H. (2002). Letters to the Times. *Los Angeles Times*, May 14, B12.

CAUGHT ON TAPE

It was eerily dark out that evening at the approach of midnight on October 1, 1996 when Captain Eric Schreiber and Copilot David Fernandez stepped outside onto the tarmac under their AeroPeru Boeing 757 jet for the preflight walkaround at Lima's Jorge Chavez International Airport. The big plane was quite a sight up above them, all clean and shiny after a 25-hour stop in the maintenance facility. The polished aluminum underside of the long fuselage hovered overhead, silhouetted by the dark and starless Lima sky. The plane sat higher up above the ground than most due to its tall landing gear, and Schreiber and Fernandez had to crane their necks back and look almost straight up during the walkaround. They would take off and rise above the marine layer at a thousand feet and head south on their regularly scheduled 3.5-hour flight to Santiago, Chile.

The preflight inspection was routine. Down low the tires and landing gear looked fine, and higher up the engine inlets were clear. Even further up, the doors and hatches were all closed and tight on the body. There were no visible irregularities on the fuselage or flight control surfaces, Pitot tubes or static ports, air conditioning inlet or antennas. They saw no drips or leaks, loose rivets or missing bolts, or any corrosion or specs of dirt, for that matter. Everything, as far as one could see that black night, appeared to be in absolutely tiptop shape. Or so it seemed.

❖❖❖

Shortly thereafter, around 12:30 a.m., with Schreiber and Fernandez in the cockpit, AeroPeru Flight 603 was pushed back from the gate and waved off by the ground crew outside. Inside the airliner were nine AeroPeru crew members and 61 passengers, a typical mix of South and North Americans and a few Europeans and British scattered about. The plane was far from full, so there was plenty of room to stretch out for any passengers who chose to do so.

Flight 603 originated in Miami, Florida much earlier that evening, but with a different crew and a different jet. An older plane, an AeroPeru 727, had actually made the trek down to Peru from the U.S. The 727's crew had retired for the night and now Captain Schreiber and Copilot Fernandez were picking up the flight and carrying the continuation passengers and new customers further south to Chile after the change of equipment in Lima.

Up on the flight deck behind the closed cabin door, Schreiber and Fernandez throttled up a bit and steered the plane with their little one-handed nose steering wheel located off to the side of the seat. They slowly drove the 757 out around past the terminal and on toward their assigned runway, number 15. Copilot Fernandez would execute the takeoff and climb out. Captain Schreiber in the left seat worked along with him, as they always did regardless of who was actually doing the flying, making all necessary changes to instrument settings and verifying the readouts on the displays and the positions of the controls.

Within minutes they arrived at the assigned runway. The crew and passengers in back were belted in and ready to go, some already dozing off for the flight down to Chile. The Lima tower had given all but the final authorization. The maze of luminescent electronic displays on the console cast a faint greenish glow on the pilots and the darkened cockpit. The runway lights outside marked a clear and straight path across the otherwise black landscape ahead, much like a featureless computerized display in a low-budget flight simulator. AeroPeru Flight 603, 757 N52AW, sat poised and almost ready to go.

"Lima Tower," called Copilot Fernandez on the radio, his own voice feeding back clearly into his headset. "AeroPeru 603, runway 15, ready for takeoff."

"AeroPeru 603," came the reply from the controller in the Lima airport tower. "Use noise attenuation. Wind calm. You are authorized to takeoff runway 15."

"One five, one five, transponder," said Schreiber, continuing the necessary checks and callouts.

"Flaps one five," stated Fernandez, his Latin voice resonating professional and clear. "Takeoff briefing complete."

"Takeoff 41," rejoined Schreiber in a particularly positive tone. Everything had gone so smoothly. They were exactly on time, 41 minutes after the hour. It should be a nice flight.

"What precision!" he remarked somewhat smugly to Fernandez, obviously pleased with their professionalism and clockwork performance. "Not even the Swiss..."

❖❖❖

After a few final checks they pushed the engine throttles forward and began their roll down the runway.

"Power's set," said Schreiber.

"Power's set," verified Fernandez.

"Eighty knots."

"Checked," responded Fernandez as they picked up speed.

"V-one," called out Schreiber as they reached their required speed for pulling back the stick. "Rotate," he called out a second later as the nose gear left the ground, and the cockpit and then the rest of the plane rose up into the inky sky. "V-two," continued Schreiber, again noting the key velocity target of the airspeed indicator.

The landing gear control was clicked into position four seconds later, and the sounds of the gear being pulled up into the plane reverberated through the cockpit.

"Gear up," stated Fernandez.

"Right. V-two plus ten," answered Schreiber, indicating their continued acceleration.

They were off and on their way.

Yes, Schreiber was right. They had attended to every detail once entering the flight deck, everything had been performed right on schedule with a precision difficult to match, even by the Swiss.

And it was all caught on tape.

❖❖❖

With over 21,000 flight hours under his belt and as AeroPeru's most experienced pilot, Eric Schreiber was fully capable of stepping through the routine but complex procedures of a takeoff and flying his plane with precision. The same was true of Copilot Fernandez in the right seat and presently at the controls. They were also more qualified than most flight deck crews in handling the unexpected — the sudden bout of bad weather, that confusing communication from a controller, or even the occasional equipment malfunction. Most importantly, they were highly experienced and well trained in the use of the cockpit displays and flying on instruments when nothing could be seen out the windows.

And like other experienced airline pilots, Schreiber also

knew the critical importance of the feedback provided by the cockpit displays, especially in an advanced aircraft such as the 757 in which the pilot is deprived of many of the immediate sensory cues of flight. There was no wind in his face to give a sense of speed, and his ears did not "pop" as they ascended in the comfortable pressurized cabin. On this dark night with the typical marine layer over Lima, all that he could possibly learn about the outside world and the aircraft's relationship to it had to come from the bank of luminescent electronic displays before him, displays which he had been trained to use and rely upon in moments like this. Unlike old mechanical displays from the past, these electronic displays were connected to and driven by the computers. And like a brain and its nervous system, the computers were linked to the outside world with miles of electrical connections and sensors, some inside the cockpit and some attached to the skin of the airframe. Just in case there were problems with the computers or electronic displays — which was extremely unlikely — there was a set of backup "traditional" displays to present altitude and other key information to the pilots. Schreiber might not be able to look out the window to tell how fast they were moving or how high they were on a night such as this, but he could glance at the displays and instantly know what he needed to guide the plane along the planned flight path: the yaw, the pitch, or the roll angle of the aircraft; their airspeed and their speed over the ground; their compass heading; and, of course, their altitude. They also had a forward radar system to alert them of trouble ahead.

Yet despite their skill and experience and the redundancies built into the avionics, there was little to prepare Schreiber and Fernandez for the events about to unfold within the confined and darkened cockpit of their 757, now rising up through the mist in an utterly featureless, blackened *ganzfeld*. The die had long since been cast. The expanse of humming electronic displays across their wall-to-wall console was about to take on a life of its own, invaded, it would seem, by spirits fixed on nothing less than annihilation of the aircraft and its human cargo.

❖❖❖

At just a hundred feet or so above the ground and rising fast into the void, Fernandez glanced at the altimeter, expecting to see it tick off their ascent. Instead, there was no change in the display despite the obvious fact that they had left the ground and were accelerating out and upward. The nose was angled up and the g forces from acceleration had pushed them slightly down and back into their seats as they pulled up from the runway.

"The altimeters are stuck," he stated crisply only six seconds after the landing gear had come up.

A scant two seconds later there was a bone-jarring interruption. The mechanical voice wind shear alarm came on suddenly. "Wind shear!" blurted from the cockpit speakers. And then it sounded twice again to drive the point home: "Wind shear! Wind shear!"

Both men scanned the primary displays, trying to understand. The pilots were not panicky in any way, but the unexpected events had certainly gotten their attention. Wind shear? How could there be wind shear? The sensors had detected a very dangerous condition, one demanding their immediate attention. But there was not supposed to be any weather up ahead, certainly not a thunderstorm. It was dead calm outside when they took off. It didn't make sense.

Nine seconds passed as they both studied the displays and control settings, absorbing whatever information seemed relevant, but also pausing a moment to try and make sense out of the quickly unfolding and unusual situation. They were also keenly aware of the possibility of wind shear given the presence of the alarm. Simultaneously, they continued their planned ascent and climb out, the only sensible thing to do given the critical moment in time and space.

"Hey! Altimeters have stuck," repeated Fernandez, who continued at the controls.

Schreiber responded quickly yet quietly, clearly having heard the first time, but simply focused on the display panel and trying to understand. "Yeah."

"All of them."

"This is really new..." said Schreiber evenly, his voice trailing off in thought and his eyes darting over the console for a number of seconds. But he picked up the verbal interplay again momentarily, knowing that they had to focus on the real business at hand, making sure they did not fly the plane into the ground. They had to maintain their velocity. "Keep V-two plus ten. V-two plus ten," he repeated, instructing Fernandez to watch the airspeed indicator and keep their speed up. They most certainly did not want to stall while diverting their attention to this problem.

Fernandez held the assigned heading and speed and monitored the displays. Looking out the window to see where they were did not help at all. There were no visual cues whatsoever. Nothing, only blackness. They would have to fly the plane entirely from what they could feel and hear and whatever the instruments might tell them. He looked out the window just to check once more, but a moment later scanned the airspeed indicator again. What? "The speed..."

"Eh?"

"The speed..." shouted Fernandez.

"What happened? Not climbing?"

"I'm climbing... But the speed?" he questioned out loud and

also to himself.

❖❖❖

It had now been only 60 seconds since their wheels left the runway. Slowly, ever so slowly, they were piecing together the situation, at least what they knew of it, but there wasn't enough information to get any sense at all of the broad picture and what was really going on. None of the altimeters were working, that was for sure. This had become obvious seconds after takeoff. Then there was the wind shear alarm just as they had cleared the airport. It had now sounded three times. Lima, Schreiber knew, never had normal rain, making the repeated wind shear warnings doubly puzzling. And now the airspeed indicators were not registering. The cockpit instruments were in effect telling them that they were still on the ground, not moving, but faced with a dangerous vertical wind shear condition up ahead which might plunge them back down to the earth. It was nothing short of bizarre. The last thing they needed now was another distraction.

The cockpit reverberated again with another jarring alarm a minute and ten seconds into the flight. This time it was the rudder ratio warning, telling them to avoid large or abrupt rudder inputs and indicating a problem with aircraft speed and sensors. Schreiber scanned the instrument panel again.

"How strange," was his calm but alert response. Like Fernandez a moment before, he was talking as much to himself as to his fellow pilot. Schreiber's inaction was not indicative of the significant effort he was applying to the problem, however. He knew that this was primarily a mental challenge, not a matter of wrestling with the controls. Other than all of the auditory alarms, the plane seemed to sound fine. And although the display indications were abnormal, the 757 *felt* like it was behaving normally. What were they to believe?

Just as when they had taken off a minute before and he realized they must maintain their airspeed and avoid a stall, he recognized now that they could not continue on their current eastward course indefinitely. There were tall mountains east of Lima and they were eventually going to run into one of them unless they did something.

"Turn to the right," he called out sharply to Fernandez. With sufficient altitude they could make a big sweeping turn away from the mountains and head out over the ocean behind them where they might be better able to size up the situation. Once things were under control, or when they at least had some idea of the source of the problem, they could head east again, line up with the runway, descend, and get back safely on the ground. It would be a big 100-mile loop from start to finish. They were obviously going to forget about going to Santiago.

Fernandez started the slow turn to the right while also continuing their ascent. Schreiber watched the displays. The altimeter began to change, but in the wrong direction!

"Well... Go up! Go up! Go up! Go up! Go up! Go up!" he shouted.

"I'm..."

"Go up! You are going down... eh, David."

"I am up, but the speed..." exclaimed Fernandez, noting that their indicated airspeed, and the direction of change in altitude, seemed all wrong.

"Yeah, but it's stuck," responded Schreiber.

The mach trim alarm bleated from the cockpit, showing that the plane was flying at high speed and that the automatic stabilizer had stopped functioning. Schreiber searched the panels again, seeking answers. "Mach trim, rudder ratio..." he said, again trying to talk through the problem. Then he noted the displayed altitude falling once again. "Go up! Go up! Go up! Go up heading one hundred."

Fernandez pulled back on the stick further and brought the plane heading around a little more. Then things appeared to settle down. The alarms stopped screaming at them and the airspeed shown on the display seemed to be reasonable. "The speed is better now," observed Fernandez after setting up on the new course.

But the moment of calm did not last. The cockpit avionics counterattacked twenty seconds later, this time with a vengeance. On came the auto throttle disconnect alarm, then the rudder ratio alarm, followed by the mach speed trim alarm once again.

"Yeah... s—!" "Rudder ratio!" shouted Schreiber, confused and upset about the reoccurring rudder ratio alarm.

Fernandez, on a hunch that shifting to an alternate signal source for the instrumentation might help matters, changed the source selector control, but it seemed to have no effect.

"S—, rudder ratio," blurted Schreiber.

"Mach trim, mach trim," said Fernandez, correctly identifying the mach speed trim alarm.

That was it. Schreiber had had enough. "Let's go to basic instruments. Everything has gone s—."

❖❖❖

In the two minutes and thirty seconds that they had been aloft, the controller at Lima Tower and the pilots had communicated just once with regard to a normal change in radio communication frequency. The situation in the cockpit had demanded far too much of their attention to do anything until now. They had not communicated with the passengers in back, and they would not do so for the remainder of the flight. There simply was not time. But it *was* time to make sure the folks on the ground knew what was going on. "Tower, AeroPeru 603," called Fernandez over the radio.

"We are in emergency," said Schreiber to Fernandez, instructing him to tell the tower that they were in the middle of an emergency.

"AeroPeru 603. Tower. Go on," came the response from Lima Tower.

"OK," said Fernandez over the radio, somewhat panicky. "We declare emergency! We have no basic instruments. No altimeter. No speedometer. We declare an emergency!"

"Roger," came the response back from Lima Tower. "Altitude?"

Fernandez was not sure what to say. What was their altitude? He sure as hell wished he knew. His response indicated as much, but also the deteriorating state of the flight. "We don't have... We have until thousand feet..." Then he gathered his thoughts and read off the current displayed altitude from the altimeter. Given everything that was going on, he didn't want to indicate that it was a certainty, so he softened the response a bit: "Approximately 1,700."

"1,700," repeated Schreiber.

"1,000 feet, roger, roger, AeroPeru 603. Confirm if you can change frequency to 119.7 to receive instructions from radar control." Fernandez's response was not particularly reassuring to the Lima Tower controller, so he just rounded safely down to an even thousand.

Fernandez switched the radio frequency to 119.7 as instructed. They would now communicate with Lima Air Traffic Control at another location. "We go to 119.7."

❖❖❖

Schreiber had pulled out the flight manual and was

studying it to better understand what to do with all the unthinkable alarm combinations. He handed the book to Fernandez, asking him to read a section while he continued to decipher the displays. In the meantime the auto throttle disconnect alarm, rudder ratio, and mach speed trim alarms had come on again.

Fernandez read from the book. "Auto throttle disconnect, rudder ratio and mach speed indicator."

"OK," responded Schreiber, waiting for more.

But who could sit and read from a manual at a time like this? Fernandez couldn't help but look up at the panel between his page turning. What he saw gave him a jolt. God! "500 feet!" he shouted, reading from the altimeter. "These a— h— from maintenance..." he yelled out, revealing his hunch about the root cause of their problems.

"What s— have they done?" screamed Schreiber.

Schreiber, as Captain, had the authority to take physical command of the plane at any time, and it was at this moment that he exercised this right, three minutes and thirty seconds after takeoff. "I have the command," he told Fernandez, as he placed his hands and feet on the controls.

Schreiber and Fernandez began to discuss the autopilot controls and the confusing indications about their status. They were interrupted by Lima Air Traffic Control who wanted to know what was going on. "603, Lima."

"We request vectors for ILS, runway one five," responded Fernandez. He was asking for vectors so that they could continue their turn out to sea and then head back east to the runway.

"Not yet, not yet," interjected Schreiber. "Let's stabilize." He wanted to address their current phase of the crisis and their presumed low altitude before moving to the next step. But the response from Lima Air Traffic Control came anyway.

"We suggest a right turn heading 330."

"Turning right, course 330," responded Fernandez to Lima Air Traffic control.

"Affirmative," came back Air Traffic Control. "And keep present altitude."

"Altitude?" said Schreiber. "We can't..."

Fernandez interjected, talking to Air Traffic Control. "What level do we have? Do we have 4,000 feet? Let's see if you confirm us."

"Correct," came the reply. "Keeping 4,000."

❖❖❖

Two minutes later AeroPeru Flight 603 was 30 miles from the airport, crossing the coastline and headed southwest on a heading of 205 degrees out over the Pacific. All airspeed indicators read zero. The altimeter showed 9,200 feet, a fact, so it appeared, just confirmed by the air traffic controller on the ground who was reading the altitude transmitted by the 757's transponder and displayed on his radar screen.

Schreiber and Fernandez struggled to stabilize their position and course prior to continuing their turn to the right. They fully retracted the flaps, which were still partially extended, and discussed the operation of the autothrottle and autopilot. Captain Schreiber wanted to engage the autopilot, but after another minute of toying with the controls and tense discussion with Fernandez he gave up and decided to fly the plane manually. They both continued to vacillate between believing and disbelieving what the instruments said. Yet they were never fully confident in their conclusion one way or the other. All the while they shot through the black sky, unable to see the stars above or the lights below.

A few short minutes later, ten minutes and 39 seconds into

the flight, Lima ATC informed them that they were now 40 miles from the city. The ATC controller's screen displayed 603's broadcast altitude of 12,000 feet. He had also calculated the speed of the jet by hand, noting its change in position over time. "AeroPeru 603, you are 40 miles from Lima and... at level one two zero [12,000 ft.]. Approximate speed over the ground is 310 knots."

"Roger," responded Schreiber. "We are at 12,000 feet and we have recovered speed now." Fernandez picked up the transmission a dozen seconds thereafter, informing Lima ATC that they were on course 220 and turning to the northerly course 330.

Schreiber instructed Fernandez to continue reading through the manual while he focused on the controls. "Read all that..."

Fernandez read aloud from the alphabetized listings in the list of contents, searching for some explanation or insight into the causes and recommended response to the constant, discordant alarms, as the instruments — both computer-driven and backup electromechanical displays — vacillated wildly and unpredictably.

"Basic instruments," shouted Schreiber, repeating himself from a few minutes earlier. "Let's go to basic instruments!" They seemed to be providing the most reliable readouts, but he was not even sure about that.

"Basic instruments," agreed Fernandez.

❖❖❖

Shortly after 13 minutes into the flight they had completed this segment of their turn and were flying on course 330. "AeroPeru 603," called out Lima ATC. "You are at 40 miles... flying parallel course 330..."

"Correct," responded Fernandez. "We request vectors from

this moment on."

"Roger. We suggest course three six zero," a heading directly north and at a right angle to the approach vector for Lima and the airport 40 miles to the right.

Communications between the cockpit and Lima ATC were interspersed with curt and tense interchanges between the pilots as the airspeed indicators continued to display conflicting and extreme values. One minute it looked like they were slowing down, and the next they thought they were speeding up, and they could not even determine if the autopilot and autothrottle were connected or disconnected.

"It's falling!" shouted Fernandez, referring to the airspeed.

"S—. Yeah!" yelled Schreiber. But moments later the airspeed indication climbed up, and continued to climb.

"It's going up too much," shouted Fernandez.

Alerts and alarms continued to reverberate within the cockpit. Lima ATC cut in again, alerting them to their position and warning them that they would have to initiate their turn eastward at some point in the closing minutes. "...You are crossing radial 230 from Lima. Distance west-southwest is 37 miles."

"Correct," responded Fernandez. "We will..." he began, but broke off and pleaded their case to ATC once again. "We have problems here reading instruments. You will have to help us in altitude and speed if possible."

"OK. Roger."

Fernandez redirected his attention to setting up the systems for capturing the Lima Airport localizer beam which they planned to use to turn toward the airport when the time arrived. For the next several minutes Schreiber struggled to make some sense of the auditory alarms, visual displays, and his own kinesthetic feelings about what the plane was doing and where they were. One second the overspeed alarm would come on, indicating dangerously excessive speed, yet moments later the stall warning fired off, his control stick shaking mechanically in his hands, warning of dangerously low airspeed and impending stall. He pulled the throttles back, but the airspeed indicators showed they were accelerating!

Lima ATC cut in again at just under 17 minutes into the flight to update them on their position. "You are crossing the 260 of Lima, at 31 miles west. Flight level is 100 plus seven hundred [10,700 feet]. Approximate speed is 280 over the ground." The controller had once again calculated their speed by hand based on the time it took them to travel from one point to the next.

"Perfect," said Schreiber to Fernandez.

Schreiber's confidence was encouraging, but not entirely justified. "Yeah," related Fernandez to Lima ATC, not quite as optimistically. "But we have an indication of 350 knots here..."

❖❖❖

Ten seconds later a two-tone overspeed alarm sounded and stayed on. It was too much for Schreiber, who moments before believed that it looked like they had a chance.

"F— s—! I have speed brakes. Everything has gone. All instruments went to s—. Everything has gone. All of them!" he shrieked.

On came the stall warning alarm. It sounded again in eight seconds and six seconds after that and then again in five seconds.

"We are going down!" screamed Fernandez.

He got on the radio with Lima ATC. "Is there any possibility..."

The Lima ATC controller was reading his mind. "Correct. Rescue has been warned..."

"We request... is there any airplane that can take off to rescue us?"

"Eh?... Wait. No, no, no, no!" shouted Schreiber to Fernandez. Schreiber had lost his composure a moment before, now Fernandez had lost his, or so it appeared. But Fernandez and the controller were thinking the same cloudy thoughts, that somehow another plane might, just might, be able to guide them in safely.

"Yes. Correct. We are going to coordinate immediately. It's being coordinated immediately," came the controller's response to Fernandez.

"Any plane that can guide us, an AeroPeru that may be around? Somebody?"

"Don't tell him anything about that!" screamed Schreiber again, wanting to face the task at hand and not willing to admit or reveal that level of desperation.

Fernandez disagreed. "Yes!" he shouted. "Because right now we are stalling..."

Lima ATC broke in. "Attention. We have a 707 that will depart to Pudaheul. We are telling him."

Schreiber looked at his airspeed indicator, which showed that they were traveling at 395 knots and in an overspeed condition, not a stall as indicated by both the ongoing stall warning and his copilot's currently low airspeed reading. "We are not stalling." he screamed. "It's fictitious. It's fictitious."

"No!" snapped Fernandez. "If we have shaker. How would it be not..."

"But even with speed brakes on we are maintaining 9,500... I don't understand... What power do we have?"

Fernandez read off the current displayed airspeed, 395 knots, and then settled down for a half-minute while they went over the instruments again, never giving up on their frantic efforts to find something they could believe, each man continuing to flop back and forth between confidence and panic, illusion and reality. And who was to say which was which?

"AeroPeru 603," radioed Lima ATC. The controller had been noting their track on the screen, the altitude transmitted to him by their transponder, but calculating their speed by hand again based on the distance traveled over a fixed time. "You have turned slightly to the left. Now you are heading 320 and your level is 100 [10,000 feet]. Approximate speed of 220 and a distance of 32 miles northwest of Lima."

His callout was interrupted by the onset of a piercing double alert alarm in the cockpit.

❖❖❖

For slightly over a minute they frantically discussed the continuing overspeed alarm, stall warning, another plane being sent up to help guide them in, and the apparent disconnect between their throttle inputs and aircraft speed. They were just about to begin their turn to the right to line up with Lima and the airport.

"Too low terrain!" blared the mechanical voice from the terrain warning system. "Too low terrain!" it began to repeat, on and on.

"What happened?" snapped Captain Schreiber.

"Too low terrain."

"But..." Schreiber knew — or thought he knew — that they were well out over the ocean, west of land and west of Lima at an altitude of 10,000 feet. He acquiesced to the inanimate voice. They must have somehow ended up over land again. "Let's go left."

Fernandez keyed the microphone again and informed Lima ATC of their intent. "We have terrain alarm. We have terrain alarm."

A cacophony of voices and alarms rose up from the bowels of the cockpit panels. "Wind shear!" "Wind shear!" "Wind shear!" "Too low terrain!"

"...Flight level one zero zero [10,000 feet], over the sea, heading a northwest course of 300," came the voice from the Lima ATC controller over their headphones.

"We have terrain alarm and we are supposed to be at 10,000 feet?" responded Fernandez to the controller.

"S—!" rejoined Schreiber to Fernandez. "We have everything!"

"Wind shear!" "Wind shear!" "Wind shear!" screeched the automated voice.

"We have all computers crazy here..." said Fernandez into the microphone.

"S—!" "What the hell have these a— h— done?" shouted Schreiber to Fernandez. His frustration was not directed at the controller.

❖❖❖

Seconds later, after another series of brief interchanges with the controller regarding their course out to sea and the low-terrain warnings, Fernandez radioed the controller wanting to know if they were ascending or descending. "We have... 370 knots. Are we descending...?"

The controller had just finished calculating the speed by hand again. "You have 200 speed, approximately."

"200...?"

"220... speed over the ground, reducing speed slightly" replied the controller.

"S—!" yelled Schreiber. "We will stall..."

"Sink rate!" "Sink rate!" "Sink rate!" "Sink rate!" came the mechanical voice, warning of a high rate of descent, an alarm consistent with low speed and Schreiber's conclusion about an imminent stall.

They pushed the throttles forward and Schreiber pulled back on the stick. The mechanical artificial horizon on the panel showed their nose lifting upward; the visceral feedback told them they were heading up higher. All the while the altimeters read 9,000 feet, not changing up or down as they made major control inputs. But regardless of the unchanging altimeters, they felt that they had avoided a disaster and that things were settling down again. It was time to turn back toward Lima and try to get back safely on the ground. If they could just get down below the marine layer they would see enough to judge their altitude and speed. A safe landing would be within reach. To hell with the damn instruments.

❖❖❖

Schreiber and Fernandez decided that everything was reasonably under control now and they could start to think once again about making the right turn for the final short leg to the airport. "We will try to intercept the ILS [landing signal]..." radioed Fernandez to the ATC controller.

"Roger, AeroPeru 603. You show level nine seven hundred [9,700 feet]," said the controller, reading the altitude broadcast to him from 603's transponder.

"That is right," replied Fernandez. Moments later he was on the radio again: "Verify the speed. It is very important."

"Correct. You are starting turn... It shows a velocity of 270 ground speed," replied Lima ATC.

They decided that a speed of 270 knots was acceptable given the circumstances. Schreiber continued to execute the turn to 70 degrees, not quite due east, as Fernandez monitored still more cockpit alarms and instrument settings. Although it felt like the situation was stabilizing, the displayed data seemed as incomprehensible as ever.

ATC came on again. "Altitude is 9,700 and the speed is 240 knots over the ground," he said, reading the altitude directly from his ATC display and the speed from his own calculations. "51 miles from Lima."

Their displayed airspeed was excessively high — almost twice what the controller was telling them. Schreiber had now cut back on the power, beginning the descent. He couldn't get his mind or his eyes off the airspeed indicators. "How can it be flying at this speed if we are going down with all the power cut off?"

"Tell me the altitude please," said Fernandez to the controller.

A caution alarm sounded again in the cockpit, followed by the mechanized cockpit voice. "Too low terrain! Too low terrain! Too low terrain..."

"Keep nine seven hundred according to presentation, sir," responded the Lima controller.

The caution alarm sounded again.

"Nine seven hundred?" questioned Fernandez, just to verify their altitude.

"Yes, correct. What is the indicated altitude on board?"

"Nine seven hundred, but it indicates too low terrain," replied Fernandez.

Schreiber now had the plane lined up for the angled descent from 9,700 feet down to the airport 50 miles distant on course 70 degrees. The nose was pointed down at about ten degrees. He looked at the airspeed indicators. They read 370 knots. At what point, he wondered, should they think about lowering the landing gear?

Fernandez was thinking the same thing. "Do we lower the gear?"

"Don't know..."

The abrupt change in their direction of movement was simultaneous with the booming crunch of the left engine and wing of the airliner bouncing off the water. The caution alarm sounded, and the mechanized voice warning continued its soulless chant: "Too low terrain. Too low terrain..."

"We are impacting water!" screamed Fernandez to Lima ATC. "Pull it up!!" he cried to Schreiber, his finger now off the mike button.

Schreiber felt the plane clearly bounce up and into the air. His hands clutched the control stick. "I have it! I have it!" His voice was unwavering. For seven seconds they vaulted back up into the black sky, Schreiber struggling to keep the wings level and nose pointed forward. But there was really nothing he could do now to change their course. The damage had been done. The 757, now 200 feet up, began to roll upside down.

"We are going to invert!" he screamed.

Just as the plane rolled completely upside down and Schreiber, firmly belted into his seat, struggled at the controls, the mechanical voice alarm caught its breath again. "Too low terrain" spilled from the speakers one last time. Then all was oddly quiet as they bounded through the night, inverted, descending back towards the surface of the sea.

"Whoop. Whoop. Pull u..."

END OF TAPE

❖❖❖

In the days and weeks following the crash of AeroPeru Flight 603 and the loss of all aboard, there was no shortage of allegations and finger pointing about the cause, especially by Peruvian officials sensitive to recent U.S. criticism of their government's oversight of aviation safety. On October 3, after having listened to a tape of Lima Air Traffic Control communications with AeroPeru 603 from the previous morning, Elsa Carrera de Escalante, the Peruvian Transport Minister, announced that computer failure appeared to have caused the accident. "It seems that there was a blockage in the computer system," said Señora Carrera. "It is not the first time that one of these planes has had this kind of fault. We have to find out why the computers went crazy." Early reports by the media fueled fears that there had been a total failure of the plane's highly computerized flight control system — or "glass cockpit" — despite the fact that no such total system failure had ever occurred in a commercial airliner. These explanations also did not account for the apparent failure of the "independent" electromechanical backup instruments with which the plane could have been flown.

Other officials voiced far different perspectives. According to the head of Brazil's airline pilots union, it was the Lima ATC controller who was responsible for the crash. He had, of course, told Schreiber and Fernandez that their altitude was 9,700 feet above sea level just moments before they plowed into the ocean. The Peruvian transportation ministry was quick to point out, however, that the controller simply relayed altitude information sent to him by Flight 603's data transponder. How was he supposed to know that the plane, some 50 miles away, was only only a few hundred feet above the water?

In the meantime, the 757 — or what was left of it — was somewhere on the bottom of the ocean. The answers to the questions were probably there as well, perhaps among the wreckage or stored in the voice and flight data recorders. Within the week the Peruvian Navy located the wreckage 55 miles northwest of Lima, more than 30 miles off shore, scattered across an area 1,200 feet wide and over a mile long at a depth of 493 feet. The fuselage was largely intact.

Members of a U.S. Navy explosives team working aboard a Peruvian recovery vessel located the sonar pingers attached to the flight data and cockpit voice recorders less than two weeks after the accident. Oceaneering Technologies, Inc. of Baltimore, specialists in deep-ocean retrieval, were contracted to assist in the recovery. They were on site within days and soon retrieved bodies and the two orange "black boxes" which were carted off by the U.S. National Transportation Safety Board for analysis. They were pulling up pieces of wreckage by the first week of November.

The crash investigators — unlike some public officials and members of the media — were thoroughly familiar with the 757's avionics and reasonably certain that it was not a catastrophic failure of the computers that brought down Flight 603. It was, in all probability, far less technical than that. They probed the wreckage with open minds, but with well-founded suspicions about the root cause. The first bits of evidence came from the live television images sent back to the ship from the remotely operated vehicle cruising along the ocean bottom, its light and camera lens scanning the aluminum skin of the airframe, searching for clues. Still, when the same pieces of wreckage were finally raised from the depths and swung on board the wet deck of the recovery ship, investigators were incredulous at what they saw.

All three left-side static ports — the sensors used to measure outside air pressure and, thus, altitude and speed — were covered with masking tape. The tape had weathered 603's half-hour flight, crash into the ocean, and a month on the bottom. The static ports and the sensors within them were the source of data for the computer-based avionics and the old-fashioned backup flight instruments. Accurate data could not be transmitted to any computers or instruments with the static ports sealed over with tape.

❖❖❖

Five AeroPeru mechanics were indicted a few months later on charges of negligent homicide in the deaths of the 70 passengers and crew. The events leading up to the night of October 1 were relatively easy to piece together. For the 25 hours proceeding Flight 603, the 757 had been in the maintenance facility at the Lima Airport. Among other things, the plane was scheduled for blade changes on the No. 2 Pratt & Whitney PW2037 engine. Mechanics decided to take the opportunity to clean the aircraft and polish the aluminum underside. To prevent contamination of the static ports and per instructions in the Boeing maintenance manual, they covered them tightly and thoroughly with masking tape. Halfway through the cleaning job the mechanics were diverted back to the work on the engine. The masking tape was never removed from the static ports and, later, not detected by Eric Schreiber and David Fernandez during the preflight walkaround the dark night of October 1.

Eleuterio Chacaliaza was found guilty of negligent homicide for his role in failing to remove the masking tape from the sensor ports and was sentenced to two years in jail. His four fellow mechanics were acquitted. The U.S. National Transportation Safety Board issued urgent recommendations to airlines to use warning flags to help ensure that sensors are unobstructed prior to flight.

On the legal front, families of 58 victims were each awarded $500,000 from AeroPeru, and families of 28 victims initiated a suit against Boeing in Miami Federal District Court claiming design and training defects. Among other things, the filing claimed that other manufacturers' airliners are delivered with red port covers for use during aircraft cleaning. Boeing settled the suit out of court for an undisclosed sum rumored to be in the many millions of dollars. AeroPeru declared bankruptcy in the Spring of 2000.

REFERENCES AND NOTES

AeroPeru 603, transcription from voice recorder, October 2, 1996, Lima, Peru [translation from Spanish to English by Jorge Jarpa] (1996). *AVweb NewsWire www site.*

AeroPeru 603, transcription from voice recorder; alarms activity from voice recorder, October 2, 1996, Lima, Peru [translation from Spanish to English by Jorge Jarpa] (1996). *AVweb NewsWire www site.*

AeroPeru in bankruptcy (2000). *Air Transport World,* 37, 7, July, 161-163.

AeroPeru pinger pinpointed (1996). *AVflash AVweb NewsWire www site,* October 14, 2, 41.

Accident description: AeroPeru N52AW (1996). *Aviation Safety Network www site.*

B757 pitot static schematic. Boeing 757 operations manual (1983). August 30.

Duct tape blamed for deadly crash of Peruvian 757 (1996). *The Seattle Times,* November 5.

Duct tape caused jet crash, NBC reports (1996). *Los Angeles Times,* November 5, A4.

Gehrke, D., Johnson, T. and May, P. (1996). Mystery shrouds 757 crash killing 70 off Peru. *The Seattle Times*, October 3.

Higdon, D. (1996). AeroPeru wreckage located, DFDR recovered. *AVweb NewsWire www site*, October 21.

Higdon, D. (1996). AeroPeru 757 static ports inopped by duct tape. *AVweb NewsWire www site*, November 11.

Higdon, D. (1997). Caught on tape: Five AeroPeru A&Ps indicted in October 757 crash. *AVweb NewsWire www site*, April 21.

Higdon, D. (1997). Damn the duct tape: Boeing sued for AeroPeru crash. *AVweb NewsWire www site*, June 23.

Johnson, G. (1996). Glass cockpit: what pilots do if 'magic' fails. *The Seattle Times*, October 9.

Ladkin, P. (1997). News and comment on the AeroPeru B757 accident; AeroPeru Flight 603, 2 October, 1996 (1996). *University of Bielefeld, Faculty of Technology www site*, February 19.

Letts, Q. (1996). Computer blamed as 70 are killed in Peru jet crash. *The Times*, October 3.

Lyman, E. (1996). 757 crashes in Pacific off Peru. *The Seattle Times*, October 2.

McKenna, J. T. (1996). Peru 757 crash probe faces technical, political hurdles. *Aviation Week & Space Technology*, October 7, 21-22.

McKenna, J. T. (1996). Blocked static ports eyed in AeroPeru 757 crash. *Aviation Week & Space Technology*, November 11, 76.

Phillips. E. H. (1996). NTSB urges change in static port covers. *Aviation Week and Space Technology*, December 2, 33.

Swaine, L. and Lane-Cummings, K. (1998). Maintenance worker held liable in crash. *AVweb NewsWire www site*, January 26.

Although all dialogue in this story is taken directly from the 30-page translated transcript from the cockpit voice recorder, major elements have been left out due to its length and complex nature.

911, MORE OR LESS

Syreeta Middleton, 13 years of age, asleep on the couch in the living room of her uncle's house in southwest Los Angeles, awoke to the smell of smoke. It was not ordinary smoke, like the smoke that wafts out of a fireplace or off a lit cigarette; this smoke smelled foul, like burning plastic or synthetic fibers. The source was immediately apparent when she opened her eyes and saw the flames just feet away. The stuffed chair, the one next to the sparking electric night-light that they had noticed earlier, was on fire. Syreeta jumped up from the couch and ran for her mother in the nearby bedroom.

"Fire!" It was a simple but powerful word, especially when shouted in the middle of the night in a house full of sleeping children, in a house with steel security bars covering each and every window. The bars were there, of course, to protect them from the dangers outside in this part of the city just east of Crenshaw and north of Martin Luther King Jr. Park. But now the danger was inside — not outside — and like a cage at the zoo the security system could prevent people from getting out as well as from getting in.

Beverly Middleton, age 36, awoke to the sound of her daughter's cry and bolted up from the bed. The entire bedroom came to life abruptly as 3-year-old Patricia Marlene, 2-year-old Donovan, and 11-month-old William were also startled awake. Three other kids were asleep down the hall in another room. Beverly's son Lawrence, age 14, was away with his uncle for the evening.

The young children in the front bedroom began to cry. Beverly and Syreeta raced back into the living room. The chair was now in full flame. They found the fire extinguisher, squeezed the handle and pointed the spray, but the flames and smoke continued to spread. The blaze was much too large for the little fire extinguisher, and the flames quickly leaped to the adjacent furniture. It was clear that the entire house was going to go up within a matter of minutes. Beverly began to panic. She continued spraying the expanding fire with the small extinguisher, but to little avail. The heat grew more intense and they backed off into the hallway.

Beverly turned to Syreeta; they had to get the kids out of the house, she said. Syreeta's three other sisters, LaToya, age 11, Melissa, age 8, and Marcella, age 6, were in the back bedroom. "Get the girls out of the back room," she instructed Syreeta, "and go across the street for help." She would stay and get the littlest ones out. Referring to her own mother who lived 14 blocks away, and seemingly aware of what was about to unfold, Beverly said her last words to Syreeta: "Mom will take care of you guys."

Syreeta ran into the back bedroom where her sisters slept, shouting at them to get up and out of their bunk beds. They had to go out the window. Her uncle had told her about the security bars on the window and the special emergency foot lever that would release the bars in case they needed to get out. *Panic hardware* is what the fire fighters called it. This was the one set of bars on the house on which the panic hardware had been installed. She slid the glass window open, then stepped on the pedal down by the floor, but it did not move. She stepped on it again, this time stomping down as hard as she could. The pedal gave way and the set of bars pivoted loose, away from the outside wall. Syreeta crawled over and through the opening and stood outside in her pajamas, now covered with soot, and pulled each of her three younger sisters out the window to safety. Together they ran across the shallow front yard to the sidewalk where the two youngest girls, Melissa and Marcella, stayed at the curb. Syreeta and LaToya ran across the street to their neighbor's house to get help. It was another small stucco house in the flat streets of Los Angeles.

❖❖❖

At 1:54 a.m., not even a minute after the neighbor across the street met the two girls at the front door, a 911 call was received by the Los Angeles Police Department dispatcher's office. The call was short and to the point.[1]

"Los Angeles Police, Operator 761," answered the operator.

"Uh, could you have the Fire Department come to 3102 9th Avenue, there's a fire, the lady's [a] neighbor across the street, could you please hurry up?"

"Is it a house or apartment?"

"Uh, huh."

"The house is on fire?"

"Uh, huh."

"Is there anybody trapped inside?"

"And she have 8 kids, I'm going to go get them out of the house."

"So there's people still inside," said the operator, seeking

[1] According to department procedure, the 911 operator should have patched the caller through to the Fire Department dispatcher when she first determined that the emergency call was for a fire. She did not do so, however, due, perhaps, to the brevity and urgency of the call. Patching the call through would have saved time, made the transfer of information more direct, and resulted in the caller's address being displayed automatically on a computer display at the fire dispatch office.

confirmation.

"Umm, I don't know, I ran in the house, the neighbor told me to call you."

"OK, I'll let them [the Fire Department] know."

"OK, thank you," said the neighbor, who then hung up.

❖❖❖

Thirty-three seconds after the neighbor called 911, the Fire Department dispatcher answered a call from the 911 operator.[2]

"91, Fire Department, can I help you?"

"Hi, 91, this is 761."

"Yeah, what's up?" came the casual reply.

"Do you have a fire at 3102 South 9th Avenue?" said the 911 operator, asking if the Fire Department had received a direct report of the fire.

"3102?"

"Uh huh."

"What's the name of the street?" questioned the Fire Department dispatcher.

"9th Avenue."

"Vine?"

"Uh huh."

"No, uh, not that we show," replied the Fire Department dispatcher, indicating that they did not show a fire at a location on Vine.

"OK, we had [a] citizen reporting a house fire with possible kids still inside."

"OK," responded the Fire Department dispatcher. "Did you get his callback number?"

"Yea," replied the 911 operator, who then read the number

---

[2] Only the call numbers — and not the names — of the 911 operator and Fire Department dispatcher were made public.

to the Fire Department dispatcher. The phone number of the person who called in the fire had registered on the 911 operator's console.

"OK," continued the Fire Department dispatcher. "You said that was 3102 Vine?"

"No, 9."

"Oh."

"As in the number 9," clarified the 911 operator.

"Oh, 9."

"Uh huh."

"Oh, OK, OK then." The Fire Department dispatcher was now clear on the street name, or so it seemed.

"OK."

"OK, thank you. Bye."

"Bye-bye."

At 1:58, more than four minutes after the neighbor called 911, the Fire Department dispatcher sent orders to a nearby fire station for units to be dispatched to the 3100 block of 9th Street.

❖❖❖

Back at the scene, where the fire was now visible from outside the house, the neighbors had gathered in the dark with Syreeta and her three sisters, waiting for the sirens and big trucks, terrified about what might be happening behind the bar-covered windows and doors of the house in the 3100 block of 9th Avenue. A Los Angeles Police Department squad car cruised by the location just by chance at 1:57 and called in a message about the fire on police radio. Two more 911 calls were placed by neighbors by 2:01, and Beverly Middleton's mother had driven over from her house 14 blocks away. The nearest fire station, all the neighbors knew, was only 6 blocks down the street.

❖❖❖

Fire trucks, lights flashing and sirens blaring, turned the corner onto the 3100 block of 9th Street at 2:01. The sirens were cut and the trucks braked to a stop. The sweeping colored lights cast an erie glow down the street of modest stucco houses. No one was outside, everything was quiet and dark, and there was not a fire anywhere in sight. A fireman radioed the dispatcher, telling her that there was not a fire at the address. The crews sat in their expensive and beautifully maintained trucks, waiting for instructions. The dispatcher dialed the callback number for the original 911 call made to the 911 operator. The phone in the house across the street from the burning house in the 3100 block of 9th Avenue rang, and rang again. An answering machine picked up and the pleasant recorded greeting asked the caller to leave a message. Two more 911 calls about the fire on 9th Avenue came in, and in a few minutes it became all too apparent what was going on. At 2:05 the firefighting units were redirected to the 3100 block of 9th Avenue, about three miles southwest of their current location in the area of Los Angeles affectionately known as Korea Town.

❖❖❖

The house was fully engulfed when the first fire companies arrived at the scene at 3102 9th Avenue at 2:08. The hoses were deployed, the nozzles opened up, and the burning house entered with some difficulty and danger. Fourteen minutes later at 2:22 the fire was extinguished. Inside, in the hallway next to the living room, lay the remains of Beverly Middleton, her son Donovan, and her daughter Patricia Marlene. The body of her infant son William was found in a bed.

❖❖❖

The Los Angeles Fire Department immediately acknowledged the very slow response time of its units to the fire and the possible role of the delay in the deaths of Beverly Middleton and three of her children. In a thorough report on the investigation into the event presented to the City Council, a large number of contributing factors were cited, including the similarity of many street names within the city, the casual attitudes of the 911 operator and fire dispatcher, the failure of the 911 operator to patch the call through to the fire dispatcher when it was first received, and the subsequent failure of the actual address to appear on the fire dispatcher's computer. The report also pointed out the importance of ongoing efforts to address the dangers of old or malfunctioning security bars installed in many crime-ridden areas. Subsequent changes to operating procedures included the requirements that verbal exchanges of addresses include a cross street and that messages be repeated word for word to help verify that they have been understood. Yet with dozens of similarly named Los Angeles city streets such as Union Avenue, Union Place, Union Street, 9th Street in San Pedro and 9th Street in Korea Town, and 9th Avenue just east of Crenshaw and north of Martin Luther King Jr. Park, there was no guarantee that it could not all happen again.

REFERENCES AND NOTES

Editorial: fourteen minutes of ineptitude. Tragic 911 response marred otherwise superb citywide performance (1995). *Los Angeles Times*, February 12, 4.

Feldman, P. (1995). House fire kills 4; girl, 13, saves 3 sisters.

Tragedy: heroine's mother, two brothers and a sister die in the blaze. Mix-up that sent firefighters to wrong address prompts inquiry. *Los Angeles Times*, January 31, 1.

Feldman, P. (1995). Errors that ended in 4 fire deaths described. Inquiry: minutes after initial mix-up sent firetrucks to wrong street, other 911 calls gave correct location, deputy chief says. But response came too late. *Los Angeles Times*, February 2, 1.

Report details dispatching snafu in fire communication: document reveals staff was given the correct address four times before trucks were sent to the wrong house miles away. The blaze killed a mother and three of her children. *Los Angeles Times*, February 8, 1.

ATM

Thanksgiving — the most American of holidays. A day to give thanks, that's for sure, but also a chance to step back and reflect on life, especially the important basic things like family, friends, and the roof over your head. A day to feast, as did the small group of colonists camped on the shores of New England who started it all on the fourth Thursday in November, 1621. Many of their small band had perished during the first brutal winter in North America, victims of starvation, disease, and unrelenting cold. But most survived, and it was appropriate and practical for them to give thanks for their good fortune and feast before the snow began to fall and the cold settled in once again. So ingrained into the American colonists was the spirit of giving thanks once each year in the fall that none other than George Washington declared a national Thanksgiving Day in November of 1789. In the next century, in 1863, Abraham Lincoln decreed it an annual national holiday.

❖❖❖

So far, it had been a good Thanksgiving for James Gallagher, 379 Thanksgivings after the first one in Plymouth. Granted, western New Jersey was not as picturesque as the shores of Massachusetts in November, but they had their share of the autumn colors and the dramatic shift in seasons that occurs throughout the Northeast. James had nothing to complain about on this Thanksgiving Day. Everyone had gotten along quite well and the food had been good. The weather had been appropriately clear and crisp. In fact, there was only one thing that he actually needed after the leftovers had been put away and the dishes dried, and that was some cash. Yes, cash. Cold, hard cash. Credit cards and checks were fine, but there was no substitute for having dollar bills in your wallet, especially with the long weekend under way and all that was planned. So at 8:45 p.m., Thanksgiving Day, the fourth Thursday of November, 1999, James Gallagher was in his car driving down the road to his local branch of Commerce Bank, the one with the 24-hour Automatic Teller Machine (ATM) on East Somerset Street near the intersection of Route 206.

❖❖❖

The big, broad, red letter "C" of the Commerce Bank logo stood out on the clean white wall of the modern building ahead, illuminated by his car's headlights. There were no other cars in the small parking lot at the bank branch when James pulled in off of East Somerset. It was nice being able to park right next to the front door. This was, after all, *America's Most Convenient Bank.*

He turned off the headlights, grabbed the keys from the ignition, and headed for the small enclosed lobby a few steps away across the walkway where the ATM machine was located at the entrance to the bank. The front wall of the entrance and the front door were made of glass. The bank itself, a few steps further in, was closed up tight, as one might expect given that it was late and today was a national holiday. As the glass door closed behind him he realized how quickly the weather had changed this year and how cold it had become outside. Having the ATM machine inside a lobby was not essential, but it made things easier and a lot more pleasant during the minute or so it took to complete a transaction, especially when the weather was unpleasant.

Although it was 8:59 p.m. and dark out, James felt comfortable and secure in the ATM lobby. It was well lit, there was no one else around, and it was, after all, Thanksgiving Day, not a day on which one might be fearful of being mugged. Raritan was a safe community and he had no concerns about his safety that evening. Besides, Commerce Bank was known for its customer service and concern for safety. They regularly provided bank customers — especially regular ATM customers such as James — with helpful tips about using ATM machines. The nine-point list was widely distributed:

- Before you approach the ATM, look around and check out the area. If anything seems suspicious, return to your car and either drive to a different ATM or use it at another time.
- Be sure to lock your car and take your keys with you — don't ever leave your car running.
- If you're using a drive-through ATM, lock your car doors.
- Get your card out before you approach the ATM. If you're making a deposit, seal the envelope before you reach the ATM.
- Make sure you close the entry door if the ATM you're using has one.
- When you have completed your transaction, put away your card, receipt and cash immediately. Look around again to check for anyone who looks suspicious.
- If you are followed by a car, drive straight to your nearest police station. If you're on foot, walk into the nearest place where people are gathered.
- Block the ATM keyboard when you're entering your private PIN number.

- If you have concerns about the security at an ATM location, call Commerce Bank's Security Department at 1-888-751-9000.

❖❖❖

His ATM card in hand, James oriented it as shown on the graphic instructions and slid it partially into the slot of the machine. The ATM sucked it in from his finger tips and sprang to life. The display on the screen asked him to enter his personal identification number (PIN). He thought for a moment. There was no need to "block the ATM keyboard while entering his private PIN," as warned by Commerce Bank; there wasn't anyone else around to look over his shoulder. He punched the number into the keypad. Then the machine needed to know what kind of transaction he wanted to make. James wanted to withdraw some cash, so he selected the "withdraw" option and, when asked by the machine, selected the amount of cash from the options listed.

He would have his cash in a moment. It was a straightforward exchange between the two of them: the machine prompting James and James responding by selecting his choice and pressing a button or two. But as he waited patiently in the glass-enclosed ATM lobby of Commerce Bank in Raritan Township, New Jersey at 8:59 p.m. on Thanksgiving evening for this mindless face of stainless steel to spit out his receipt and cash, another player lurked in the shadows. It was not a mugger or thief, not a snooping passerby looking over his shoulder for his personal identification number. It was another computer, this one programmed to lock the door to the ATM lobby at precisely 9:00 p.m.

No one — someone at Commerce Bank had reasoned — would want or need to conduct an ATM transaction

Thanksgiving night, a time when all good Americans would be in their homes, watching football on the television and having that second piece of pumpkin pie. No one except James Gallagher.

The silent electronic clock somewhere struck 9:00. A line or two of software code instructed a circuit to be closed, and the electromagnet in the door lock of the ATM lobby fired. A bolt slid into place within the lock and door frame. James heard a funny little "click" behind him just as the ATM machine handed him his money and his receipt. "Thank you," it said on the screen.

❖❖❖

The sound was unmistakable. The door had just locked behind him. James took his cash, his receipt, and ATM bank card out of the machine and stuffed them into his pocket. He turned to the door and pulled on the handle. It was locked tight.

He gave the door another hard pull, and then a push, but to no avail. He turned around and looked behind him into the dimly lit bank. No one was there. Turning back to the door to the outside, he thought there had to be a handle or release switch somewhere, and he examined the door and the surfaces surrounding it. There was not a trigger on the door or grab handle, no switch or latch on the door frame, no levers or latches up at the top or down along the bottom. Nothing. There was no way to physically unlatch the lock from inside the lobby and no warning of any kind before the lock had automatically fired. Incredible!

Well, there had to be a telephone, he reasoned. He scanned the little room and the wall on which the ATM machine was mounted. There were no doors or small compartments, and there most definitely was not a telephone with which to call

Commerce Bank's Security Department on their toll-free number. In that case, there had to be a voice intercom or an alarm button, he surmised, and he searched the lobby again, all the while realizing with building anxiety that there did not seem to be any way to contact anyone outside his little glass cell. He yelled out, first restrained but then louder, but there was no one around in the bank or the darkened parking lot to hear him. He could make out the headlights of the occasional car driving down East Somerset.

And unfortunately, this part of East Somerset Street, like so many urban business areas in America, was not what you would call "pedestrian friendly." It was actually State Route 626, one of the main roads through Raritan. During the day it got its share of traffic, but one would not expect there to be pedestrians mulling about now, a little after 9:00 p.m., Thanksgiving night.

He focused, trying to think of other approaches and of what MacGyver might do. Perhaps within one of those indentations at the top of the ATM machine was a lens for a video camera. That's how they would know he was trapped! They would see him. But on second thought, any video system would only be linked to a recorder of some kind. There would not be anyone sitting behind a console looking at all these ATM customers in New Jersey, Pennsylvania, Delaware, and Manhattan. If there were a video camera it most certainly would be making only a record on a tape or a disk drive for use later in case there was a disputed transaction or unauthorized use of an ATM card. No, this was not going to help him, but it might make an entertaining piece on some reality television program, him pacing back and forth, sniffing out every corner and every surface of his glass cage.

He studied the door and its mechanism again, but came to the same conclusion as before. There was no way to unlatch the lock or even get the door off its hinges, and he had no tools or other objects in his pocket that would be of any help. This was a bank, after all.

He might be able to break the glass. But then again, there wasn't any furniture to throw against it, nothing he could use to smash through. Maybe he could lunge at it with his shoulder, pushing off from the opposite wall and giving it all he had. But what if the glass broke? He could cut himself to pieces. The glass was very thick anyway, and it didn't look like he could break it even if he tried.

❖❖❖

James yelled out and pounded on the door hoping that someone might hear him, sat down on the floor and grumbled to no one in particular, and then jumped up to see if there was some way of signaling someone through the ATM machine. He put his ATM card in the machine and went through all of the menus, searching all of the screens for an input option that might be of help. He made another cash withdraw thinking that this might alert someone or trigger a distant alarm in a computer center, but it didn't seem to do anything. He began to worry. Just how long was the ATM lobby going to be closed, anyway? Thanksgiving was a four-day holiday for most people. Was the bank going to open up again tomorrow, Friday, or would it be closed for four days, until Monday morning? Monday morning! Now that was a long, long time away. And then there was the other matter of a personal nature. Just how long could he hold out?

No point in worrying about things when you have no control over them. It was now more than obvious to James that he had no control over this mess. He sat down on the floor once again and curled up on the carpet of his glass cell. He had many hours in which to think, to decide what it was he was going to do once he finally got out.

❖❖❖

A few minutes before 6:00 a.m., Friday morning, November 26, just as the sun crested the minor hills of Raritan Township and illuminated the remaining golden leaves on the trees, a car pulled into the parking lot of the Commerce Bank on East Somerset Street. It was the assistant branch manager, arriving for the start of another banking day. With the door to his cell — and then the door to the bank — unlocked, James Gallagher made a beeline for the men's room, completed an important call on the telephone, and promptly withdrew all of his money and closed his account.

Bank spokesman David Flaherty later described James Gallagher's experience as a "very freak accident."

REFERENCES AND NOTES

Associated Press (1999). Customer spends nine hours trapped in bank's ATM lobby. *Los Angeles Times*, November 27, A23.

Commerce Bank. ATM Safety Tips (2001). *Commerce Bank www site.*

Although widely published, the central character's name has been changed in this story. All other story elements are believed to be accurately depicted.

UNDER THE RADAR

In the maintenance hangar at the Tri-State Airport in Huntington, West Virginia the evening of January 6, 2003, mechanic Brian Zias worked the night shift. The aircraft waiting to be serviced was N233YV, a 19-passenger Beech 1900D turboprop operated by Air Midwest on behalf of US Airways Express. It was due for a D-6 (Detail Six), a regularly scheduled maintenance stop in which key parts were inspected and important fluids topped off after every 100 hours of flight. These planes made money only when they were in the air and full of paying passengers, so the maintenance service was conducted at night when the demand for the aircraft was low.

After its arrival in Huntington on a scheduled flight a few hours earlier, the Beech 1900D had been moved away from the passenger terminal and pushed into the small hangar with the tail aimed toward the far back corner of the building. The hangar doors were slid closed just a few feet from the aircraft's nose. Three portable kerosene space heaters were positioned near the plane to bring up the temperature and create a comfortable working environment for the aircraft maintenance crew working the night shift.

Brian Zias had been hired by SMART (Structural, Modification and Repair Technicians, Inc.) of Edgewater, Florida not quite two months before, just a few months after the Huntington facility had opened for business. US Airways Express, a wholly owned subsidiary of US Airways, had a contract with Air Midwest (a wholly owned subsidiary of Mesa Air Group) to operate the Beech 1900D and others like it in their fleet for US Airways Express flights in and around the Carolinas. Air Midwest, in turn, contracted with Raytheon Aerospace LLC (74 percent owned by Veritas Capital, Inc.) to maintain its planes flown under the US Airways Express banner, and Raytheon Aerospace contracted with SMART to provide mechanics to perform the actual work.

To the uninformed observer these complex linkages and multiple layers of companies might appear to increase operating costs, but the opposite was actually true, especially in terms of aircraft maintenance. Layered subcontracting was a way to avoid the high costs associated with large maintenance centers and heavily staffed unionized workforces with restrictive work practices and hefty pensions.

The supervisor and trainer on site for the night shift was George States, an employee of Raytheon Aerospace. Like all of the other SMART mechanics there in Huntington, Brian Zias had no formal maintenance training or experience on the Beech 1900D, and it was up to George to oversee Brian's work and sign off on his training. George States also served as the inspector for all of the work that night and put his name on the official paperwork, but the ultimate responsibility for maintenance rested in the hands of the aircraft operator, Air Midwest. A half-dozen personnel were on hand that night in the hangar, and George States divvied up the tasks so that everything would be done by morning when the sun rose and the plane had to be towed out and put back into service.

George States assigned Brian Zias the task of inspecting the elevator and rudder flight control cables on the Beech 1900D, to make sure that they were not frayed or hanging up on anything as they ran from the two control columns in the cockpit, under the floor of the cabin, and back to the rudder and horizontal elevators at the top of the tail. The cables, rudder, and elevators were the means by which the pilot made the plane go left and right and up and down. Pushing the control column forward pulled a cable and moved the elevator down and made the nose of the plane go downward when it was in flight. Pulling the column back pulled another cable which moved the elevator upward and made the plane's nose move up. Brian was to visually inspect the cables by opening various panels, measure cable tensions with an instrument, and make any necessary adjustments for length, position, and tension, a task he had never performed on the Beech 1900D. It was an opportunity, however, to learn how things were done and get signed off on the procedure as part of his ongoing OJT (on-the-job training). Everyone has to start somewhere.

George States was a busy man that night. In addition to planning the work and overseeing everything, he was keeping an eye on Brian Zias and the work and on-the-job training of another mechanic assigned to check the play in the elevator tab system. George was also supervising the replacement of a fuel control on the number one engine, personally conducting a borescope inspection on the same engine, and serving as the final inspector for all of the work that was to be done as part of the D-6 service.

As is the case with almost every aircraft maintenance task performed by any mechanic, Brian Zias began the elevator cable inspection and adjustment task by opening the Maintenance Performance Manual supplied by Raytheon Electronic Publication System for the Beech 1900D. Brian's first step was simple enough: measure the outside air temperature near the captain's window. It was 55 degrees Fahrenheit. Page 203 of the Beechcraft *Maintenance Manual 27-30-02* displayed the Elevator Cable Tension Graph, and Brian located the current temperature and its associated cable tension on the other axis. Just as a car's tire pressure changes as it goes from cold to hot, Brian knew that the desired tension of the cables would be different depending on the ambient temperature around the aircraft, and the tension had to reflect the temperature.

He referred back to his paperwork which instructed him to inspect the elevator, trim tab, and rudder cables and measure their tensions using a cable tensiometer. The cables ran under the floor of the cockpit and cabin, back to the tail assembly, and up to the moving parts of the tail. They were accessed by removing panels on the floor of the cockpit and a panel on the floor of the rearmost baggage compartment at the back of the plane. The readings from the cable tensiometer showed that the tensions on the rudder and tab cables were fine, but the elevator cable tensions were low. Brian turned back to *Elevator Control — Maintenance Practices in Maintenance Manual 27-30-02* and began the procedure to tension the elevator cables.

The first step was to "Disconnect the autopilot servo cables." This aircraft did not have an autopilot, so he proceeded on. Step two instructed him to "Locate and remove the flight compartment seats...," but Brian could tell that there was absolutely no need to remove the seats as he had already accessed the cable through the panel in the cockpit floor. This specific procedure was obviously for another model of Beech aircraft, a suspicion confirmed when he came to the third step instructing him to "Locate and remove the passenger seats, carpet, and floorboards on the right side of the passenger compartment to gain access to the elevator cable turnbuckles" — which would be a huge job. He had already accessed the cable turnbuckles through the panel inside the aft luggage compartment back near the empennage, and there was certainly no need to pull up the passenger compartment floorboards. Again, this procedure was clearly for a different model.

Step four instructed him to "Adjust the center-to-center

length of the push-pull tube assembly between the control column and the forward elevator bellcrank to a dimension of 15.12 ± 0.06 inches." Once again, it did not look like this step applied based on the visible hardware, and Brian asked George States for his opinion. George looked it over and agreed that the step was not required.

Step five instructed him to "Adjust the surface stop bolts on the elevator control horn support for 'up' travel of 20° ±1° and 'down' travel of 14° ±1°." Brian decided that this really did not require taking a complicated measurement of the elevator control horn support mechanism and that it would be far simpler to crawl up the rolling ladder near the tail and manually move the elevators up and down to see if the elevator was hitting the stops. Sure enough, the elevators moved up and down just fine when he lifted and lowered them, although he could not see into the cockpit while standing up at the tail and be certain that the control column was also moving through its full range of motion.

As with a number of the earlier steps, steps six, seven, and eight did not seem to be necessary and appeared to apply to a complete overhaul or re-rigging of the system rather than the simple cable tension adjustment he was performing. Step six told him to "Verify that the bob weight stop bolt clearance is 0.5 ±0.06 in.," step seven was to "Adjust the forward bellcrank stops for 0.37 ± 0.06 in. clearance from the stop bolts," and step eight was to "Verify the forward bellcrank stop bolts make contact before the bob weight stop bolts make contact with the weight." Brian Zias talked these steps over with George States and they agreed that they applied to a far more extensive rigging overhaul and that none of them were required during this D-6 service.

Brian was finally to the point where he could adjust the tensions on the elevator cables. He crawled back into the aft baggage compartment and over to the open hatch above the cables and their respective turnbuckles. Each turnbuckle was about the size of a large finger with the cable attached to either end. Turning half of the turnbuckle one way shortened the cable and turning it the opposite way lengthened the cable. One cable and turnbuckle was for "up" elevator control and the other was for "down" elevator control. Step nine read "Remove the safety clips from the turnbuckles and release cable tension." Brian removed the clips, turned each turnbuckle to release the tension on the rigging, attached the cable tensiometer, and turned the turnbuckles to tighten the cables. Each turn of each turnbuckle rotated a threaded bolt and shortened — and subsequently tightened — the rigging. He finished with the turnbuckles and reset the safety clips to hold them in position. Finally, step ten called for a "Control Column Support Roller Inspection" to verify that, among other things, the range of motion of the control column in the cockpit moved in step with the movement of the elevators and also that the neutral position of the control column was the neutral position of the elevators. But as before, it seemed that this was a complex procedure that was associated with a complete re-rigging rather than the simple retensioning being done here. Brian Zias and George States talked things over once again and agreed that this last step did not apply to their procedure. After a few more minor activities the elevator cable adjustment service was recorded on the Maintenance Record for Work Order #05-1030106016, form #M001, page 3 of 5, item 18. The record read "Elevator cable tension low" and part number "MS21256-2." Under the "Nature of Action" heading, the following was written: "Adjusted elevator cable tension per BMM 27-30-02. Ops check normal."

It had been a long night.

❖❖❖

At the US Airways check-in counter in the Charlotte-Douglas International Airport terminal in Charlotte, North Carolina a day later on the morning of January 8, 2003, the US Airways agent greeted each passenger checking in for an upcoming departure. The US Airways Express flight to Greenville-Spartanburg, South Carolina was scheduled for 8:30 and the passengers were arriving with plenty of time to spare. The aircraft, N233YV, a 19-passenger Beech 1900D operated by Air Midwest on behalf of US Airways Express, was already at the airport and would be ready to leave on time. It was going to be a full flight, and based on the luggage being checked in and the carry-ons being lugged and rolled onward to the gate, the cargo hold and the overhead bins would be filled to capacity. The airline made money only when the planes were in the air and full of paying passengers, so the check-in agent was pleased to do the work and get the passengers processed.

There was nothing particularly unusual about the 19 individuals arriving at the counter for the 8:30 flight to Greenville-Spartanburg. They were a representative mix of business people and leisure travelers with a handful of college students thrown in. There was also one child. But to anyone who had an observant eye and a decent memory and had worked the check-in counter or gate for a decade or two and could think back and compare all these passengers to their counterparts of twenty years before, these people looked big. Yes, big. This group of passengers as a whole wasn't really any different from, say, a planeload of people traveling to Raleigh or Asheville. But in terms of how this group or any similar sample of people might have looked ten or twenty years before, the differences were striking. People, especially people in the Southeast, were just larger than they used to be. Nevertheless, the photo IDs were inspected, seats assigned, boarding passes issued, and checked bags handed over. Routing tags were printed and attached to handles. Luggage was lifted and plopped onto the moving conveyer behind the counter and disappeared behind the rubber curtain as passengers shuffled off over to security screening and their gates beyond with their coats and briefcases and purses and backpacks and carry-ons in tow.

❖❖❖

On the ramp outside the gate next to Air Midwest N233YV, the gate agent and baggage handler (an employee of Piedmont Aviation subcontracted by Air Midwest to handle the baggage on its flights for US Airways Express, the subsidiary of US Airways) lifted the checked bags off the carts into the aft cargo compartment at the back of the airplane, counting the bags as they went. Although it was actually only a single compartment, the rear cargo hold had two sections — aft and forward — which were divided by a cargo net. There was also a small cargo area in the forward part of the cabin up toward the front of the plane on the right side, but that was usually reserved for use by the crew.

There was a bit of confusion about the actual number of bags, but when all was said and done it looked like there were 23 checked pieces of luggage and a tire. Two bags were particularly heavy and seemed to weigh 70 to 80 pounds each. It took the baggage handler and the gate agent working together to lift them into the cargo hold. The passengers, in the meantime, were boarding the plane up the short stairway and finding their seats on either side of the narrow aisle. The carry-on items that would not fit under the small seats or in the modest overhead compartments had been tagged and left on the ramp near the stairway and were to be loaded into the cargo compartment as well. They had not even started to bring over the carry-on items when it became apparent that everything was going to have to be very tightly packed if all the cargo was going to fit in the most forward of the two parts of the rear cargo compartment, as was usually the case. The final count on the bags was 31 — 23 checked bags and another 8 carry-ons full of clothes, toilet kits and extra shoes, laptops and new batteries, cell phone chargers, compact cameras, and PDAs. Finally, though, they got it all in and managed to make it fit like a three-dimensional puzzle. But the plane looked heavy, especially tail-heavy, as if it was sitting up a bit with its tail squatting down on the ground and its nose pointed up into the air. In fact, the tail was so low and the nose so high that the front landing gear extension was fully extended.

❖❖❖

On the flight deck of Air Midwest N233YV sitting on the ramp of the Charlotte-Douglas International Airport in Charlotte, North Carolina at 8:20 in the morning, Captain Katie Leslie was in the left seat and her first officer, Jonathan Gibbs, was in the right. Leslie and Gibbs had flown N233YV on six legs the day before, finishing up at Charlotte-Douglas International at 8:45 in the evening. Another crew had taken over from there that night, flying the plane up to Lynchburg, Virginia and then back down to Charlotte-Douglas International the next morning, today, January 8. Each of the nine flights of N233YV since the D-6 maintenance in Huntington the night of January 6 had been with light to moderate loads with an average of only five passengers. Leslie and Gibbs were starting out their day of short hops with a full load of 19 passengers (18 adults and one child) and their possessions. The airline made money only when the planes were in the air and full of paying customers, so the pilots were pleased to do their jobs and get the maximum load of passengers to their destination.

Despite being only 25 years old, Katie Leslie had accumulated 2,800 in-flight hours and much experience in the Beech 1900D — 1,800 hours, in fact. She was by all accounts a motivated and accomplished pilot. An honors graduate of Louisiana Tech University in 1999, she stayed on at the University to serve as a flight instructor before being hired by Air Midwest. Jonathan Gibbs, 26, her first officer, had logged 700 hours on the aircraft type and was also versed and skilled in its operation.

The passengers were all now on board and the checked bags were being loaded into the forward section of the aft cargo compartment. This was a small aircraft on a short hop without a flight attendant. Captain Leslie stepped out of the cockpit into the cabin aisle and greeted the passengers as she had done hundreds of times.

"Good morning, welcome aboard US Airways Express service to Greenville-Spartanburg. It's a very short flight, maybe thirty minutes gate to gate. We ask you to keep your seatbelts buckled till we're at the gate. Anything you brought with you needs to be stowed underneath your seat for takeoff and landing. We have two emergency exits on the left, one on the right. To open those doors pull the handle down, turn the door sideways, throw the door out and run out."

"This is your emergency briefing card," she continued, holding up a sample card for all to see. "It's in the seat in front of you. We're gonna play a briefing. Please pay attention as we taxi out. This door," she said, pointing to the remaining hatch, "can also be used as an emergency exit. Push the button in the box, lift the handle, and the door will come out. Please don't hang on to that door. If you do it'll pull you out. Sit back, relax, enjoy the flight and we'll have you there in a few minutes."

With the audio tape inserted and the recorded safety briefing ready to play, she returned to the cockpit and stepped over into her seat on the left. For the next three minutes captain Leslie and first officer Gibbs talked about the heavily loaded plane and worked their weight and center-of-gravity calculations.

Leslie asked Gibbs if it was clear out his right window in preparation for starting the starboard engine: "Clear over there?"

"Capped and clear on the right," came the response.

They continued with the prestart checks when the baggage handler, who was obviously concerned about the load, yelled up to captain Leslie through her window: "How many we gotta take off?"

"We're figuring it out," she called back. "We don't think we have to take anything [off]."

Leslie turned back to Gibbs, who was still doing the weight calculations. "I didn't know we were gonna be nineteen and overloaded or I wouldn't have let you... I mean... not let you... do all this," she said apologetically, referring to the 19 passengers and the full load of cargo.

"No. I got ya."

"I don't really care... how fast or slow you go," continued Leslie.

"...four... eight... one... zero... seven. Cool... seventeen oh eighteen," said Gibbs to himself as he added up the numbers for the weight calculation on the load manifest.

With 19 passengers, 23 checked bags, 8 carry-ons, a tire, and their own luggage and fuel they were right up against the limit of 17,120 lbs according to the Air Midwest estimating procedures. The aircraft's operating weight was 10,673 lbs. The 19 passengers weighed an average of 175 lbs each in winter clothing according to FAA guidelines and the reference table on the load manifest, for a total of 3,325 lbs. The coat rack contained an estimated 10 lbs. The forward cargo hold aft held

31 bags and, according to the table on the load manifest, contained 775 lbs based on the FAA's assumption that the average bag weighed 25 lbs. The aft cargo hold aft held a tire, estimated to weigh 45 lbs. The baggage handler had moved all of the luggage up into the forward half of the aft baggage hold to get the center of gravity as forward as possible, although it was still biased toward the tail. The fuel weighed 12,200 lbs, for a grand total of 17,028 lbs. Gibbs mistakenly added and wrote down 17,018 for the weight computation on the load manifest, a 10-lb error that was of no consequence given the maximum gross takeoff weight of 17,120 lbs for the Beech 1900D. They were about 100 lbs under the limit according to the charts and the FAA estimates, and it was nearly time to start up.

"Seventeen one twenty is our weight, huh?" said captain Leslie.

"Yes, is our max."

"So we're cool."

"So yeah."

At 8:29 the bags were all loaded and the weight calculations complete. The gate agent pulled the chocks from the wheels. Leslie, noticing some odd looks by the gate agent and baggage handler outside, laughed, as did Gibbs.

"He's probably looking at our tail like... 'bout ready to hit the ground right now, with all the bags back there'... laughing at us."

They worked through their checklist, with Gibbs reading the item and Leslie verifying.

"After start checklist avionics master?"

"On."

"Engine anti-ice?"

"On."

"AC buses?"

"On."

"EFIS aux power?"

"On."

"EFIS power?"

"On."

She hit the switch to play the safety briefing tape for the passengers, finished up the final checks with Gibbs, made sure everything was clear again and throttled up and guided the plane out away from the gate.

"Welcome aboard," came the recorded voice in the cabin. "The flight crew is making final preparation for departure. As they do, please note that the Beechcraft nineteen hundred airliner has many features for your comfort and safety. As these features are presented, please follow along with a passenger information card which can be found in a seat pocket..."

❖❖❖

From the cockpit of the American Airlines MD-80 jet on the ramp at the Charlotte-Douglas International Airport at 8:31 in the morning of January 8, 2003, the first officer could not help but notice Air Midwest N233YV, Flight 5481, a Beech 1900D. This pilot had, after all, 2,000 hours of stick time in the turboprop in prior years and was very experienced in its operation. It was, she said, "heavily, heavily loaded... tail-low and nose-high." Other pilots on the ramp and taxiways also could not help but take note of the appearance of N233YV, including one who saw the way the "main landing gear struts were compressed, the tires were compressed and the nose strut was almost fully extended."

❖❖❖

By 8:33 they had taxied over to the hold line to await their turn to take off. Commercial jets, cargo planes, and corporate jets came and went, and Leslie and Gibbs casually discussed what each would be like to fly while they waited a few minutes to hear from the tower. Katie especially liked the looks of the CRJ Challenger, a small but new high-end corporate jet. "Wish I was flyin it," she laughed. "I would love to be captain on that because... you decide you hate the airlines, then you've got that Challenger rating already."

"Oh yeah... yeah, wouldn't that be cool?" agreed Gibbs.

"Yeah. I mean, I might have to, you know... deal with dinners in Paris."

"Yeah."

"Or overnights in Cancun."

"You might suffer through it," quipped Gibbs, as they both broke out laughing at the fantasy.

As the parade of aircraft continued by, Katie Leslie and Jonathan Gibbs periodically stepped through a series of checks of displays and control positions and carried on with the small talk while they waited for final clearance from the tower. Another CRJ Challenger jet, the same model Katie Leslie joked about flying to Paris for dinner, rolled by and Gibbs picked up his cue. "Gosh, that sure is a nice-looking plane," knowing that this would get a rise out of his captain. "I should shut up," he laughed. "I'm doing to you right now what you were doing to me with the Krispy Kremes yesterday," a reference to Katie tempting him with one of the tasty but calorie-laden doughnuts the day before during a break between flights in the crew lounge. Leslie and Gibbs were young, fit, and not at all overweight, and each could well afford to eat a doughnut now and then.

"At least I got to eat one," she laughed.

"Jim," referring to one of their colleagues who was with them yesterday in the crew room, "Jim was probably looking for any excuse he could find to eat one."

"Sometimes I secretly go get them at the grocery store."

"Oh yeah? I've been known to do that from time to time," confessed Gibbs as they both laughed again.

At that moment, at 8:45 and 25 seconds in the morning, the tower chimed in: "Air Midwest fifty four eighty one, runway one eight right, taxi into position and hold."

Gibbs responded to the tower: "Position and hold runway one eight right Air Midwest fifty four eighty one."

Gibbs checked the transponder's mode and other readings. Leslie checked the exterior lights, and each confirmed the clearances on their respective sides of the plane. Not one to let the moment get away, Leslie picked it up again while waiting for the final word from the tower. "I love those Krispy Kreme doughnuts with the... the icing filling on the inside."

"I was just saying...," continued Gibbs, "I started to say before, you need to get a box of those and wrap them up in wedding gift wrap paper and take them to Jim's wedding."

They laughed at their friendly — and seemingly private — joke.

"That's a good call. I'm gonna do that."

"A huge box, you know, like the size box... that was in the crew room yesterday."

"That is an awesome idea."

"Wouldn't that be great?"

"Yeah," laughed the captain.

At this moment at 8:46 and 18 seconds the tower cut in. "Air Midwest fifty four eighty one... turn right heading two three zero, cleared for takeoff."

Gibbs replied to the controller. "Two three zero cleared for takeoff, Air Midwest fifty four eighty one." "Two thirty, cleared to go," he said to Leslie.

They pushed the throttles in the center console forward.

"Set takeoff power please."

"Power is set," said Gibbs.

The small but powerful turboprop airliner accelerated quickly off its mark and in 8 seconds was at 80 knots. "V-one, V-R," called out Gibbs at their first two critical speed marks, with "V-two" coming three seconds later, noting the point at which she could pull back on the "stick."

Captain Katie Leslie, thoroughly focused on the task at hand, pulled back on the two-handed flight control that stuck out from the cockpit panel and Air Midwest N233YV lifted into the air. The lever was moved and the landing gear started to come up.

She knew within a second that something was not right. The nose seemed to lift up excessively just as the landing gear started to retract. Leslie pushed the controller forward to compensate and bring the nose down and keep a conservative angle of climb with maximum lift as they accelerated upward from the runway. Nothing happened. She pushed further and harder on the control to pull the cable that ran under the floor of the cockpit and back to the tail and up to the horizontal elevators to make them move downward and lift the tail and put the nose back down, but the plane's angle of ascent continued to increase as they accelerated and moved up from the runway.

"Wuh. Oh." "Help me," she barked, pushing on the control to no effect.

Gibbs' hands held tightly on his flight controller on the right side of the cockpit, and he pushed hard, then again as hard as he could. The nose continued upward and the plane's angle increased from ten to twenty to thirty degrees.

"You got it?" called out Leslie.

"Oh."

Their full breaths and desperate grunts came involuntarily as they pushed with every ounce of life on their paired controls, four seconds of heavy full breaths and primal sounds picked up by the cockpit voice recorder.

"Push down," called Gibbs, and he took another breath. "Oh."

A half-second later N233YV was angled upward at 58 degrees, her nose pointed to the sky as if in a hammerhead maneuver at 10,000 feet at an airshow. Their forward movement over the runway fell rapidly and the plane lost lift due to the excessive angle of ascent.

"You uh," spilled out captain Katie Leslie's mouth just before the stall warning horn sounded.

The pushing and grunting continued as they vaulted unwillingly up into the sky. "Push the nose down," shouted Katie Leslie above the din of the twin turboprops and Gibbs' desperate exertions at his control. Yet neither control moved any further forward than it had when she first tried to bring the nose back down after they left the runway seconds before. The die was cast two nights before and their fate sealed when the full load of passengers and their bags came onto the plane. They were at the end of the range of control movement and nothing could have moved the stick further forward. "Ahh" came a loud cry. Then "Oh my God!"

The Beech 1900D slowed to 31 knots forward speed only a third of the way down the runway and pointed 68 degrees upward toward space as it reached the apex of its rapid climb at an altitude of 2,090 feet. It seemed to hang for a moment before rolling over to the left then on over to its back and began to fall to earth like a fluttering leaf.

Leslie keyed the radio, but it was quite hopeless. "We have an emergency for Air Midwest fifty-four eighty-one."

A child's voice from the passenger cabin was picked up on the cockpit's voice recorder. "Daddy!"

Captain Katie Leslie hung on. "Pull the power back."

The turboprop rolled again nearly 180 degrees and was now right side up but falling fast toward the airport — a small airliner full of people, their many possessions, and a load of fuel.

"Oh my God. Ahhh!"

"Uh Uh..."

❖❖❖

The ensuing investigation of the crash and loss of all aboard US Airways Express Flight 5481 by the National Transportation Safety Board was as multifaceted as the causes of the accident. The most important element was, of course, the improper tensioning of the elevator cables on the Beech 1900D during the D-6 maintenance service at the hangar in Huntington the night of January 6 and the associated reduction in the range of motion to the control column. NTSB investigators determined that the cable and associated turnbuckle for the nose-down elevator rigging was out of position by nearly two inches, the consequence being that a full two-thirds of the normal range of motion of the pitch-down control was eliminated. The pilots had pushed the flight control to its full forward position, yet the elevator at the top of the tail was, in effect, at only one-third of its down position. The problem had not been detected prior to the morning of January 8 because the loads of all previous flights had been relatively light, and there had not been a need to apply major pitch-down inputs.

The second direct contributor was the heavy load and rearward center of gravity of N233YV during Flight 5481. The pilots, using FAA tables to calculate total passenger, luggage, and carry-on weights, underestimated the total takeoff weight by 600 pounds. Evidence of the ever-increasing body weight of Americans had existed for decades, but the FAA continued to permit the use of outdated weight calculation charts out of general inaction and concern for reducing maximum passenger loads on small commercial flights. As a consequence of NTSB's investigation of the January 8 crash, the FAA instructed regional operators of 10-to-19 passenger airplanes to conduct weight surveys of passengers and their possessions. The findings were unequivocal — and sobering. US Airways Express sampled 3,018 passengers on their flights and found that the average weight, including clothes and shoes, was 200 lbs. The average weight for the entire sample collected by over a dozen airlines during the spring of 2003 was 195.6 lbs., yet the FAA-approved weight charts used by the Flight 5481 crew were based on an average wintertime weight of 175 and an average summertime weight of 170. The average weight of carry-on items for a sample of 1,538 items from US Airways Express in the spring of 2003 was 20 lbs, and 15.7 lbs for all of the airlines surveyed, yet the FAA's weight estimate for carry-ons had been 10 lbs. Checked bags had been estimated to weigh an average of 25 lbs, whereas the sample of 2,510 bags conducted by US Airways Express found a mean bag weight of 30 lbs, not far off the average of 28.8 lbs for the larger sample from all of the participating airlines nationwide. Flight 5481 was technically within weight limits based on the estimating techniques, but most certainly over the aircraft's allowable takeoff weight. And although the baggage handler and gate agent had filled the forward portion of the aft cargo hold with luggage, while keeping the rearmost portion relatively empty, the aircraft's center of gravity was seriously aft of the center of lift and sufficient to pull the tail down and nose up. The pilots, however, could not compensate due to the restricted movement of the control resulting from the maintenance work on the cables.

The layers of corporations and the prevalence of third-party maintenance outsourcing by regional carriers also came under scrutiny in the investigation, as did the fact that the FAA operation certificate for the Huntington facility was issued to Air

Midwest, yet the facility was run by employees of Raytheon Aviation LLC who, in turn, supervised employees of SMART who had no formal training on the aircraft type. The facility and its operations fell under the jurisdiction of the FAA regional inspector.

Raytheon Aviation LLC and Air Midwest noted that the cable tensioning procedure in the Beechcraft maintenance manual was vague in parts and that many of the steps were taken from other aircraft and did not apply to the 1900D. Beechcraft, in turn, responded that the mechanics skipped numerous steps in the procedure and that, if properly performed, would have taken hours more labor than was actually consumed. The FAA noted the impropriety of an aircraft maintenance supervisor also performing the final inspection on the maintainer's work, violating industry practices of independent verification of maintenance before returning an aircraft back to service.

In the end, SMART transferred Brian Zias to another facility, Raytheon Aerospace LLC relocated George States, Air Midwest revised the cable rigging procedure in the maintenance manual and added a manager to the night shift, Louisiana Tech University established the Katie Leslie Scholarship for Professional Aviation, the FAA revised its average weight assumptions and aircraft balance guidelines, and an NTSB member stated that the maintenance outsourcing system and use of third-party mechanics had flown "under the radar" of the FAA and its regional maintenance inspector.

REFERENCES AND NOTES

Aarons, R. N. (2003). Cause and circumstance: Charlotte crash. *International Aviation Safety Association www site*, June 25.

Adair, B. (2003). Unclear manual may have contributed to fatal air crash. *St. Petersburg Times*, May 26.

Aircraft maintenance and records group factual report, accident DCA03MA022 (2003). Washington, D.C.: National Transportation Safety Board, March 12.

Aircraft performance group, crash site factual report, accident DCA03MA022 (2003). Washington, D.C.: National Transportation Safety Board, March 25.

Alexander, A. and Reed, T. (2003). NTSB wraps up hearing into crash. *The Charlotte Observer*, May 22.

Alonso-Zaldivar, R. (2003). Airlines ordered to survey weight on small planes. *Los Angeles Times*, January 28, A10.

Attachment III, (accident) DCA03MA022 DFDR plotted data (2003). Washington, D.C.: National Transportation Safety Board.

Attachment 5 to Operational Factors / Human Performance Group Chairman's factual report, accident DCA03MA022: Load manifest and worksheets from Flight 5481 (2003). Washington, D.C.: National Transportation Safety Board.

Attachment 21 to Operational Factors / Human Performance Group Chairman's factual report, accident DCA03MA022: Weight and balance control programs for 10 to 19 seat airplanes operated under 14 CFR 121 (2003). Washington, D.C.: National Transportation Safety Board, January 27.

Attachment twenty-seven, maintenance sign-off paperwork, accident DCA03MA022 (2003). Washington, D.C.: National

Transportation Safety Board.

Attachment twenty-three, Detail Six paperwork (D-6 inspection procedures checklist), accident DCA03MA022 (2003). Washington, D.C.: National Transportation Safety Board.

Beech 1900D airliner maintenance manual (UE-1 and after): Aileron control rigging - maintenance practices (2002). Raytheon Electronic Publication System.

Chairman's factual report of investigation, cockpit voice recorder (transcript), accident DCA03MA022, docket No. SA-523, Exhibit No. 12-A (2003). Washington, D.C.: National Transportation Safety Board, April 17.

The Charlotte Beech 1900D crash: N.C. crash investigators eye maintenance (2003). *International Aviation Safety Association www site,* May 21.

FAA changes weight rules after N.C. crash (2003). *United Transportation Union www site,* July.

Group chairman's factual report, operational factors group, accident DCA03MA022 (2003). Washington, D.C.: National Transportation Safety Board, March 27.

Human performance specialist's factual report, accident DCA03MA022, Docket no. SA-523, Exhibit No. 14-A (2003). Washington, D.C.: National Transportation Safety Board, April 25.

Investigation update: Crash of Air Midwest Flight 5481, Charlotte, North Carolina (2003). Washington, D.C.: National

Transportation Safety Board, January 28.

Loss of Pitch control caused fatal airliner crash in Charlotte, North Carolina last year [press release] (2003). Washington, D.C.: National Transportation Safety Board, February 26.

Lowe, P. (2003). Misrigged elevator and tail-heavy condition spelled doom for 1900D. *Aviation International News www site,* July.

Pasztor, A (2004). Report on 2003 US Airways crash rebukes FAA for lax oversight. *The Wall Street Journal,* February 27, B4.

Pasztor, A (2003). Crash probe reveals maintenance gaps. *The Wall Street Journal,* May 21, D3.

Polek, G. (2003). Air Midwest crash ushers in regionals' day of disaster. *Aviation International News www site,* February 2003.

Power, S. (2003). Passengers will be weighed before boarding 19-seat planes. *The Wall Street Journal,* January 28, D4.

Power, S. (2003). FAA ponders adding 10 pounds to fliers' estimated weight. *The Wall Street Journal,* May 1, B1.

Press release: NTSB to convene hearing May 20 on Air Midwest flight 5481 crash (2003). Washington, D.C.: National Transportation Safety Board, April 24.

Public hearing on Air Midwest flight 5481: Agenda (2003). Washington, D.C.: National Transportation Safety Board, May 20-21.

Reed, T, Alexander, A. and Lunan, C. (2003). Crash puts spotlight on contracts. *The Charlotte Observer*, March 16.

Report of aviation accident, loss of pitch control during takeoff, Air Midwest Flight 5481, Raytheon (Beechcraft) 1900D, N233YV, Charlotte, North Carolina, January 8, 2003, NTSB/AAR-04/01; public meeting of February 26, 2004 (2004). Washington, D.C.: National Transportation Safety Board, February 26.

Specialist's factual report of investigation, digital flight data recorder, accident DCA03MA022 (2003). Washington, D.C.: National Transportation Safety Board, March 27.

Systems group chairman's factual report, accident DCA03MA022 (2003). Washington, D.C.: National Transportation Safety Board, March 21.

Systems group chairman factual report addendum, aircraft control cable temperatures, accident DCA03MA022 (2003). Washington, D.C.: National Transportation Safety Board, April 17.

Systems group chairman factual report addendum #2, aircraft control cable length, accident DCA03MA022 (2003). Washington, D.C.: National Transportation Safety Board, April 17.

Note: The pseudonym "Jim" is used for the Air Midwest crewman discussed during the cockpit conversation.

SAFER THAN SAFE

It had taken more than a year to plan, execute, and analyze the data from the largest pharmaceutical field trial ever conducted, and today, April 12, 1955, the tenth anniversary of the death of President Franklin D. Roosevelt, Dr. Thomas Francis of the University of Michigan was about to announce the results to an anxious public at the carefully orchestrated press conference in Ann Arbor. To anyone familiar with the world of science, this was not a typical setting in which to present the findings from a groundbreaking experiment — the largest and certainly one of the most controversial medical experiments of the twentieth century. There was not the usual sea of straight-laced and well-scrubbed medical researchers before him, name tags pinned to their lapels and conference proceedings in hand. In their seats, mainly in the front rows of the auditorium, were disheveled reporters and fidgety cameramen by the score, not one of whom sat still as Dr. Francis approached the podium in elegant Rackham Hall in the center of the university's campus. Sixteen newsreel and television cameras mounted atop a platform built above the last rows of seats in the back panned the sweeping stage. This gathering had more to do with "the top of the news" than peer review and follow-on research. This audience was here to get the scoop, file the report, and find the catchiest headline or film clip on this story — the hottest news story of the year. The people were dying to hear the outcome.

Although the audience did not know the official results

which were to be delivered momentarily, everyone present and nearly every American knew what this show was all about. One word — polio — summed it up, and one man — Jonas Salk — who was now on stage with Dr. Francis and a small panel of speakers, was synonymous with the effort to rid the world of this devastating neuromuscular disease. Since 1947 young Dr. Salk and the University of Pittsburgh Virus Research Laboratory had worked incessantly to develop an effective vaccine. Funding for the research came largely from the National Foundation for Infantile Paralysis and had been raised through the March of Dimes program. The public regulators, in an era in which *any* federal involvement in health care smacked of "socialized medicine," had stayed at arm's length from the undertaking. Yet after the collection of all those truckloads of dimes and other funds, after years of painstaking work in Pittsburgh and elsewhere, the Salk vaccine, as it had come to be known, had been put to the test by the National Foundation. The year before, in 1954, the parents of over 400,000 American school children had agreed to let their sons and daughters participate. Half had been injected with the experimental vaccine in three stages and the other half received injections of a harmless placebo. Neither the administrators nor evaluators knew who had been given the experimental Salk vaccine and who had been given the placebo, a classic double-blind study. Most scientists agreed that the trial had been well designed and executed. They could expect to have a high degree of confidence in the outcome.

Dr. Francis took the podium before the nervous crowd and spoke for one hour and thirty-eight minutes in a calm and unemotional voice, keeping strictly to the facts. With supporting charts and graphs from an overhead projector, he walked the audience through the design of the study, the organizations involved, the research methods, and, finally, the results. The outcome was unequivocal: The vaccine worked. There were three main types of polio virus: Type 1, by far the strongest strain, Type 2 and Type 3. The Salk vaccine had been 60 to 70 percent effective against the more virulent Type 1 paralysis, Dr. Francis explained. For Type 2 and Type 3, the vaccine had been even more effective — over 90 percent compared to the control group. And the vaccine was 94 percent effective against the rare but dreaded bulbar polio. Like icing on the cake, Dr. Francis also reported that there had not been a single case of vaccine-induced polio among the hundreds of thousands of children inoculated during the field trial. The Salk vaccine was deemed safe and effective, and everyone present, to say nothing of the millions of Americans listening in by live radio and television, breathed a collective sigh of relief.

❖❖❖

When the scientific presentation adjourned at two o'clock, Salk and the four other speakers went upstairs to a separate press conference where they faced the reporters' questions directly. A number of the inquiries concerned the safety of the vaccine. Wasn't it true, they said, that the vaccine actually contained the polio virus, although it was supposedly killed? Could the scientists guarantee that there was no risk involved? And what, it was asked, of the possibility that something might happen during the manufacturing process, the chance of a step being left out or a slight change to the concoction? Was it at all possible that vaccine might contain some of the live polio virus, or, worse, that a person could actually contract polio from the vaccine?

The response was confident and persuasive. Salk's answer that afternoon, as well as a few days thereafter when talking to the press, was clear, succinct, and true to form for this scientist who always looked so comfortable and assured when appearing before the media. "Massive safety tests in three separate laboratories guarantee that any live virus will be discovered and the whole batch destroyed. That is the purpose of the tests." The field trial had gone on without a hitch. There had been absolutely no cases of vaccine-induced polio in the hundreds of thousands of inoculations. The process used to kill the live virus was effective and the integrity of the vaccine batches was confirmed with rigorous tests. The process was foolproof. "The vaccine is safe," said Jonas Salk in response to the reporters' questions on April 12, "and you can't get safer than safe."

❖❖❖

A few hours later in another orchestrated event in Washington, D.C., Oveta Culp Hobby, the Secretary of Health, Education, and Welfare, was waiting to sign the documents to license six pharmaceutical companies to manufacture the Salk polio vaccine for nationwide distribution. The proceedings were highly unusual and a marked departure from the typical months-long licensing process. Although she had had nothing to say about the development of the vaccine during the previous two years, there sat Mrs. Hobby before the assembled phalanx of reporters and television cameras in Washington. A direct telephone line linked her to Ann Arbor, where a committee of fifteen virologists, including Salk, had been assembled to give their licensing approval to the government just hours after the day's presentations on the results of the field trial. Once the committee of scientists gave the go-ahead, Secretary Hobby would sign the fast-track licensing documents before the television cameras and the manufacturers could release the product.

Each scientist on the committee had been handed Dr.

Francis' 150-page report on the field trial just hours before and had not been able to read it through due to the hectic pace of the day. Such things took time to digest and discuss, and there were a few protests on the other end of the line in Ann Arbor about being put on the spot like this. The scientists should not be forced to give their blessing just because the Secretary and television cameras were waiting in Washington. They would simply have to wait, the committee decided, and for the next two hours the members talked about the results of the trial, vaccine production and testing reliability, the rush to full-scale vaccinations, and the disturbing note that batches of vaccine containing live virus had been detected (and destroyed) during the field trial.

Some even pointed out the potential difficulties of high-volume commercial manufacturing of the vaccine, where many people, new procedures, and large amounts of materials would be involved. The vaccine used in the field trial had been made in relatively small batches with great attention to every detail, they noted. Each and every lot had been tested and retested. Could the same standards be applied consistently on a commercial scale? The virologist responsible for full-scale manufacturing responded by saying that each manufacturer's production protocols — their paperwork — would be inspected and approved. As far as testing each and every bottle of vaccine was concerned, it was not possible — and unnecessary. After all, the plan was to inoculate an astounding 57 million Americans by midsummer. Vaccine quality could be assured by verifying that the manufacturers were following the production protocol which had proven to be effective in the field trial.

Salk had said that he actually wanted another year to work on the vaccine. His lab had improved its strength during the year in which the field trial was under way, and he also wanted to run more tests on the optimum spacing between the second and third shots of the three-shot inoculation regime. Immunity was known to build slowly, but at what rate no one was certain. Preliminary results suggested that it should be more than the spacing used in the field trial. Other scientists, especially Albert Sabin, felt that Salk's entire approach was flawed and that an inactivated live virus oral vaccine — rather than the dead virus vaccine developed by Salk — was the only way to provide lasting immunity to the polio virus. An oral vaccine, once perfected, would also be infinitely easier to administer.

No committee member could deny, however, the undercurrent and single reason for the entire effort in the first place. Like the filling streams and rivers at the end of winter and the onset of spring, polio season was fast approaching. It was a warm-weather disease, contracted by children from playmates in swimming pools and sandboxes. In the past half-dozen summers polio had wreaked havoc on the nation, especially children and their grieving parents. Three years before, in 1952, more than 20,000 Americans contracted the disease in its various forms. Hundreds died, thousands were severely crippled, and for many the only hope of survival was being sealed up to the neck in an iron lung. None of this would change if they did not move forward. They had a vaccine, it had been proven safe and effective, and in the end, two hours after beginning their discussions, the committee gave a unanimous recommendation to Secretary Hobby to license the Salk vaccine for manufacture. In Washington at the other end of the open telephone line she signed the papers in front of the few remaining reporters. The manufacturers had their marching orders.

❖❖❖

Before the start of the 1954 field trial, the National

Foundation approached 10 respected drug manufacturers to determine their interest in producing the Salk polio vaccine. Each firm was well-established and thought capable of undertaking the demanding manufacturing task. Six companies responded, and each was subsequently licensed to manufacture Salk vaccine for the field trial: Parke-Davis, Eli Lilly, Sharpe and Dohme, Pitman-Moore, Wyeth, and Cutter Laboratories of Berkeley, California. The production methods for Salk vaccine were, to say the very least, unusual and complex, and each of the six manufacturers put up the sizable investment in facilities and equipment necessary to start production. But when, in the early spring of 1954, it was time to begin the inoculation of children for the field trial, only two manufacturers — Parke-Davis and Eli Lilly — had produced vaccine lots that were ready for use. Production lots from the other manufacturers were not yet complete or arriving too late to use in the field trial, as was the case for three lots manufactured by Cutter Laboratories. Accordingly, only the lots provided by Parke-Davis and Eli Lilly were used for inoculations in the field trial.

The manufacturing procedures employed by Parke-Davis and Eli Lilly for the field trial were precise and controlled, but also labor intensive and expensive, more like a scaled-up version of the laboratory procedures created originally by Salk than a full-blown industrial process designed to make enough vaccine to inoculate 57 million people three times over. Every effort had been taken to maintain the large margin of safety Salk established during laboratory production. The vaccine was brewed in small batches under the watchful eyes of highly trained and experienced biologists and chemists, and, unlike many other aspects of the polio eradication program, the federal government was very much involved at this stage; government biologists, physicians, and virologists monitored the manufacturing steps and products at the two pharmaceutical firms. The small production lots meant also that the production process was relatively continuous; batches in process were not left to stand and sit for any length of time because things moved along quickly. Each lot was tested as it progressed and also when it was completed. Eleven consecutive batches of perfect vaccine had to be produced or the process was shut down, the vaccine thrown out, everything cleaned up, and the whole thing started up again. The personnel knew more about the process, the product, and the related principles of biology and chemistry than any other people on the planet. Certainly no one wanted any defective vaccine getting out, especially as part of the all-important field trial. As stated by Salk on April 12, 1955, at the conclusion of the field trial, the lots of vaccine had been checked and checked again.

The process of manufacturing Salk polio vaccine was as macabre as it was unique. It began, curiously, in India, where rhesus monkeys from the north were captured by paid hunters, caged, and carried on foot to distant train stations. From there, they were shipped by rail to Delhi, placed on aircraft and flown 4,000 miles to London where they made their connecting flights on to New York on the other side of the Atlantic, another trip of 3,000 miles. Trucks took them from New York to the central monkey processing and distribution center, Okatie Farms in South Carolina, 700 miles to the south. The six vaccine manufacturers and assorted research labs and universities around the country placed their orders with Okatie Farms, who shipped them the animals.

The number of monkeys required was staggering by any measure. All told, the vaccine manufacturers placed orders for more than 200,000 animals in preparation for commercial production for the 1955 season, a demand for laboratory primates that caused a serious rift between the governments of India and the U.S. Although devout Hindus with knowledge of the eventual fate of the monkeys refused to participate, the trapping of monkeys continued. The largest order — for 69,800 monkeys — was placed by Eli Lilly. The smallest order — for 5,200 monkeys — was placed by Cutter Laboratories in Berkeley. Each of the six manufacturers, in addition to other laboratories using monkeys for polio research, built massive facilities to house the animals after they were shipped to them from South Carolina and until it was time for each to meet his or her fate in the name of public health.

Unlike a bacterium which can be grown in a nutrient compound, polio is a virus and requires living tissue for its own replication. Large quantities of the virus were needed to manufacture the massive quantities of vaccine, and, fortunately for humans but unfortunately for the monkeys, Salk and others had determined through research that monkey kidney cells were ideal for growing polio virus. Male monkey testicles were also found to be a good medium for growing the virus, but they were smaller and the supply would be half that of monkey kidneys. So kidneys it was. Once housed in each manufacturer's laboratory, monkeys were, as needed, taken from their cages and anesthetized. A surgeon removed the kidneys and killed the animal with an overdose of ether. Their kidneys were cut into small pieces, mixed with a special nutrient compound developed by Salk, and rocked by machine for six days to stimulate cell growth. Live polio virus was introduced into each bottle which was rocked again for another four days. The virus invaded the monkey kidney cells and replicated. At this point there was a thousand times more polio virus than originally introduced. This material was transferred to larger 21-gallon bottles and chilled for a period of time, leaving a lethal preparation of live polio virus suspended in solution. Formalin — a liquid form of formaldehyde — was introduced next to kill the virus. This process was performed separately for the three strains of the virus: Types 1, 2, and 3. The three batches were filtered to remove the kidney cells and then mixed together to form one final batch of vaccine. These major steps, plus hundreds of smaller steps along the way, required weeks of painstaking labor.

The vaccine licensing documents signed by Secretary Hobby stipulated that each manufacturer follow the protocol. The original manufacturing instructions were written by Salk, but edited and added to by others. Key sections of the document were highly specific. The amount of Formalin to be added to a batch to kill the virus, various heating and chilling times, and other steps in preparation were spelled out as in a chef's detailed recipe, giving some credence to Salk's nemesis, Albert Sabin, when he said that Salk's vaccine was the product of "kitchen science." Other sections of the protocol were somewhat general, however, like a recipe that allows a cook to blend ingredients or spice a dish to personal taste. Manufacturers had to demonstrate "a consistency of performance," said Salk, but the manner in which this critical requirement was to be met was not defined. Whereas a recipe might tell the cook to remove the pan from the heat for ten minutes before the next step, the time between many steps in the vaccine protocol was left unspecified. The laboratory manufacturing protocol had been a continuous, low-volume operation performed in a lab under tight constraints where variations in timing had not been an issue. Much was not known about the production methods that would be developed for industrial-scale manufacturing of the vaccine, so certain things were left unsaid or not described in detail. But as any experienced chief knows, cooking for an army is not the same as cooking for a dinner party, and it was inevitable, given the limited time allowed and the overall lack of experience in industrial-scale manufacture of biologicals, that the outcome was not quite what everyone expected.

❖❖❖

Of the six companies involved, Cutter Laboratories of Berkeley, California was to manufacture the smallest percentage of vaccine for the national immunization effort in the spring of 1955, yet their total volume was still sizable. Cutter's production output was to be injected into roughly ten percent of the 57 million people slated for inoculation in 1955. Cutter, founded in 1914 and now run by the Cutter brothers, made biologicals, blood fractions, pharmaceuticals, hospital solutions, and intravenous equipment. The Cutter Laboratories — with a half-dozen smoke stacks rising into the sky and an assortment of large and small buildings erected over the years in whatever space was available — looked much like a typical manufacturing factory of the era. A spreading lawn, an American flag, and a giant "CUTTER Laboratories" sign spelled out in neon lights adorned the face of the property in front of the main offices and largest production buildings.

At Cutter, as well as at the five other manufacturing facilities, production of vaccine was left up to the company as long as they stayed within the guidelines of the protocol. Everyone at the National Foundation, the government, and the laboratories assumed that the basic production methods would be the same for the six manufacturers. The layout of the plant, the flow of the operation, the volume of containers, and the size of individual batches reflected individual company practices, experience, and available equipment, however. Vaccine was produced in lots which varied in size from 40 to 400 liters depending on the manufacturer. Eli Lilly and Parke-Davis, having been responsible for the vaccine used during the field trial, had experience making vaccine under strict guidelines that had, in effect, proven effective based on the successful outcome of the trial experiment. As in the field trial, these companies tested each batch of vaccine to ensure that it did not contain live polio virus. The four other manufacturers, including Cutter, did not have the benefit of this experience and therefore proceeded to develop their systems based on the protocol and in-house know-how.

The complex laboratory-scale production methods for the Salk vaccine were not at all compatible with industrial-scale manufacture. Tight time constraints imposed by the national immunization program did not make matters any easier. It was not only, as might be implied by the Salk vaccine protocol, a matter of "scaling up" what was done on the table tops of a lab. The kidneys of thousands of monkeys, not just a dozen, had to be removed and handled; massive glass bottles and tanks of biological material heated, cooled, shaken, treated, and transported; giant batches of solution containing live polio virus inactivated, filtered, and stirred, all under strict environmental and sterile conditions.

At Cutter the new large-scale operation occasionally backed up. Vaccine lots sat idle for days until the equipment required for the next step in the production line was free to use. In some instances, cooking the vaccine solution to inactivate live polio virus at specific stages was done at the exact temperatures and number of days prescribed in the protocol — but the cooked tanks of material were a few days older and therefore more potent than the material used in previous small-scale production in Salk's lab. Filtration procedures were slightly different and mechanical stirring techniques were not exactly the same as those used in the lab. Most critically, tanks of cells and live virus awaiting inactivation by Formalin sat for days, not hours, allowing kidney cells to gather in microscopic clumps that greatly reduced the effectiveness of the Formalin, allowing live virus to survive. The total result of the industrial production methods at Cutter was *tolerance stack-up*, in which the combined effect of a number of small and seemingly independent system variations renders an entire system ineffective. Unbeknownst to anyone at Cutter Laboratories — or anyone in the Department of Health, Education, and Welfare — two of the nine lots manufactured for the national vaccination program by Cutter Laboratories contained measurable amounts of live polio virus. But all of their procedures, all of their steps, all of their cooking times and measurements were within the specifications laid out by the vaccine manufacturing protocol written originally by Jonas Salk and approved by the U.S. government.

The final barrier to catastrophe was testing and measurement, and the production system failed again on a number of counts. First, the 1955 manufacturing license issued by the government dropped the original "eleven consecutive good-batch" rule from the production procedure used by Parke-Davis and Eli Lilly to greatly speed up the process. Second, whereas every batch of vaccine leaving a plant used in the field trial was tested for the presence of live polio virus, manufacturers were now permitted to simply ensure "a consistency of performance" using their own methods of choice. Some, like Cutter, adopted sampling methods in which only random batches were tested; others, like Parke-Davis and Eli Lilly, continued using the comprehensive quality testing approach from the field trial. Third, tests to detect live polio virus were used at the limits of their sensitivity. In fact, most batches of Salk vaccine contained some live virus, usually in infinitesimally small and inconsequential amounts. Tissue culture tests were not as sensitive in the factory as they had been within the tight controls of the laboratory, however, and batches containing high levels of live virus were tested and approved for release. Furthermore, higher-sensitivity tests could be conducted with monkeys, but this was expensive and very time-consuming, so manufacturers often opted for less-expensive and, unknown to them, less-sensitive culture tests. Fifth, the government's only source of knowledge of results — other than any change in the incidence of polio among the population days or weeks after inoculation — was the laboratories' production paperwork sent to the National Institutes of Health, demonstrating that they had followed the production protocol to the letter during the manufacture of every lot. Tests on all final product — the lots of vaccine soon to be injected into millions of American children — were not required and not performed by Cutter. The paperwork — not the final product — was verified by the government regulators on the other side of the continent. And as further evidence of the lack of oversight and communication between all involved, unknown to the management at Cutter, other manufacturers had produced batches containing live virus. However, these batches were subjected to more comprehensive testing and they were destroyed before being distributed.

So confident was Cutter Laboratories in their process and the quality of their product that they offered free polio vaccinations to all employees, shareholders, and their families. The manufacturing process had gone without a hitch; no contamination had been detected in any of their tested batches. The protocol was followed and the verifying paperwork submitted and approved by the National Institutes of Health.

❖❖❖

Eight of the nine lots of vaccine, including the two contaminated lots, were distributed to locations in California, Idaho, Arizona, Nevada, New Mexico, and Hawaii. Between April 12 and April 27, 308,000 first- and second-graders plus 82,000 patients at private medical offices were given their first inoculation in the three-injection series from the two contaminated production lots. Just as batches of polio vaccine might contain amounts of live polio virus (albeit in miniscule counts), each individual had a predisposition or susceptibility to the virus based on the level of existing antibodies developed naturally. Most people had relatively strong existing natural immunity, some had moderate immunity, and a smaller portion had little or no natural immunity. Like two slightly overlapping bell-shaped curves in which the right tail of the left overlaps the left tail of the right, a percentage of all injections were about to infect a percentage of people with the disease. Whereas most individuals inoculated with the contaminated vaccine fought off the virus naturally or developed only a mild temporary reaction, 10 to 25 percent of the nearly 400,000 individuals were now technically infected. A few hundred of these people were in grave danger of developing full-blown polio and losing their lives.

The first hints of trouble came in a telephone call from Idaho to the U.S. Public Health Service on April 26. Six children who had received the Salk vaccine a few days before had come down with polio. Some questioned if the children had contracted the disease prior to receiving the inoculation, but the evidence to the contrary was convincing: polio was essentially nonexistent in Idaho this time of year, and, most disturbingly, paralysis began in the injected arm — not in the lower extremities as was most often the case. A report of another five cases arrived by noon the next day, this time from California, and officials there halted all inoculations. The affected children in California and Idaho had received vaccine manufactured by Cutter Laboratories of Berkeley, California. The U.S. Surgeon General Dr. Leonard Scheele immediately ordered all Cutter vaccine recalled, and two scientists were dispatched to Berkeley to undertake an investigation. Within two days a new Poliomyelitis Surveillance Unit had been formed in Atlanta to monitor the disease on a national basis, and a select group of scientists, including Jonas Salk, were assembled to advise the government. On May 8th, with five million children having been inoculated, Surgeon General Scheele ordered the entire polio vaccination program shut down until things could be sorted out.

❖❖❖

The investigating scientists at Cutter began their work on April 28. Other specialists were brought in to examine ventilation, sanitary factors, the engineering of the plant, and other possible causes. The initial findings were particularly disturbing, not for what they found but for what they did not find — nothing. Everything was in order, the facilities were sanitary and professionally run, the equipment in good shape, the personnel competent, and the protocol had been followed. Yet two batches of the nine that had been made at the plant were now known to contain live polio virus. The answers came slowly but were clear in the end.

The combined effect of a number of relatively small "changes" in the use of the protocol in industrial-scale production was to generally increase the presence of live virus and reduce the chance, ever so slightly, of it being killed during treatment with heat or Formalin. This, coupled with a new understanding of the questionable sensitivity of the tests and testing strategy practiced at Cutter Laboratories, explained how tainted vaccine had been distributed. Members of the team traveled from Berkeley to other manufacturers' facilities across the country where they found more lots of vaccine containing live virus. Fortunately, none of it had been distributed for use on the public.

And what of the new vaccine-induced cases of polio? The final impact was tragic. Ninety-four children developed polio directly from the Cutter vaccine. All were from the Type 1 virus, the particularly tenacious Mahoney strain. And, in something not seen before, "satellite cases" of the disease sprang up quickly among family and community contacts, including 126 parents and siblings of the inoculated children and 40 members of the community with whom they had come in contact. In all, 240 cases of polio were known to have been caused directly or indirectly by Cutter vaccine. Three-quarters of these people were paralyzed and eleven died.

❖❖❖

On May 15, after having a reasonable understanding of the problem, the Surgeon General ordered the resumption of the vaccination program. Unfortunately, public confidence in the whole scheme had been undermined and the lines of people waiting to be inoculated fell to half their previous length. The summer of 1955 was warm, particularly in New England, and polio cases skyrocketed into the thousands, largely due to the hesitancy of so many millions of people to receive the vaccine injections, despite the latest assurances that it was safe. But with a far more specific production protocol and changes to tests and testing, the Salk vaccine was indeed now safe, and within a few short years the disease was all but eradicated.

The *Cutter Incident,* as it came to be known, changed the course of American history in a number of ways. The tremendous success of the Poliomyelitis Surveillance Unit in Atlanta, the first organized epidemiology study unit in the country, was instrumental in the establishment of the Centers for Disease Control and set new standards for monitoring disease outbreaks. Cutter Laboratories spent the next five years in litigation and eventually sold out to another firm. San Francisco attorney Melvin Belli, subsequently dubbed the "King of torts," launched his career on the case and acquired large settlements for some of the polio victims. In court, Cutter was found to have actually followed Salk's vaccine production protocol, but still found liable under the concept of "implied warranty," representing a major shift in American product liability law. Within a few short years Albert Sabin had developed the oral polio vaccine which was adopted worldwide due to the ease with which it could be manufactured and administered.

REFERENCES AND NOTES

Bayly, M. B. (1956). The story of the Salk anti-poliomyelitis vaccine. *WHALE www site*, Nov. 2000.

Blume, S. and Geesink, I. (2000). A brief history of polio vaccines. *Science*, 288, 1593-1594.

Cutter spokesman comments (1955). *The New York Times*, November 10, 19.

Hero with something to prove (1993). *Los Angeles Times*, March 7, A1.

Hinman, A. R. and Thacker, S. B. (1995). Invited commentary on "The Cutter incident: poliomyelitis following formaldehyde-inactivated poliovirus vaccination in the United States during the spring of 1955. II. Relationship of poliomyelitis to Cutter vaccine. *American Journal of Epidemiology*, 142, 2, July 15, 107-108.

Klein, A. K. (1972). *Trial by fury: the polio vaccine controversy*. New York: Charles Schribner's Sons.

Neal, N. and Langmuir, A. D. (1963). The Cutter incident: poliomyelitis following formaldehyde-inactivated poliovirus vaccination in the United States during the spring of 1955. *American Journal of Hygiene*, 78, 29-60.

Offit, P.A. (2005). *The Cutter incident: how America's first polio vaccine led to the growing vaccine crisis*. New Haven: Yale University Press.

Paul. J. R. (1971). *A history of poliomyelitis*. New Haven: Yale University Press.

Poliomyelitis vaccine, part 1. Hearings before the Committee on Interstate and Foreign Commerce, United States House of Representatives May 25 and 27 (1955). Washington, D.C.: United States Government Printing Office.

Poliomyelitis vaccine, part 2. Hearings before the Committee on Interstate and Foreign Commerce, United States House of Representatives June 22 and 23 (1955). Washington, D.C.: United States Government Printing Office.

Poliomyelitis vaccine, part 1. Hearings before the Committee on Labor and Public Welfare, United States Senate, Eighty-Fourth Congress, First Session, on S. 1984 and S. 2147, June 14 and 15 (1955). Washington, D.C.: United States Government Printing Office.

Scheele letter and part of report on Salk Vaccine Program (1955). *The New York Times*, June 10, 12.

Spector, B. (1980). The great Salk vaccine mess. *Antioch Review*, 38, Summer, 291-303.

Smith, Jane. S. (1990). *Patenting the sun: polio and the Salk vaccine.* New York: William Morrow and Company, Inc.

U.S. lays defects in Polio Program to mass output (1955). *The New York Times,* June 10, 1.

RHYMES AND REASONS

"Far out!" And was it ever. Henry John Deutschendorf, Jr., age 53, in typical fashion, did not bridle his enthusiasm about the glistening new LongEZ experimental plane on the tarmac at the Santa Maria Airport in California. It was the perfect aircraft for a man born in Roswell, New Mexico, of all places, the son of a U.S. Air Force test pilot. The plane was audacious, personal, and, well, a real spaceship. John, as he preferred to be called to avoid confusion with Henry John Sr., like his father had flown a lot of airplanes through the years, but this exotic number might turn out to be the best of all — fast, definitely eye-catching, and a blast to fly. Dad would have been proud.

New? No, not exactly. The LongEZ wasn't actually new, but the fresh paint job by the local refurbishing shop shined as white as the sun on new mountain snow and looked as slippery as a sheet of melting ice. Subtle multicolor accents on the wing-mounted cargo pods made a classy finishing touch — not so small as to go unnoticed and not so large that they detracted from the striking futuristic form of the Y-shaped composite airframe sitting on the pavement in the full morning sun. This LongEZ looked like an instrument sculpted by segmented porcelain insects on a distant planet, designed to propel an earthly being or two through the atmosphere at breakneck speed for over 1,000 miles, a craft he had wanted to fly and own more than any other. And own it he did. His new FAA call sign was up on the tail where it belonged, painted in crisp block letters. It read N555JD. Yep, there was simply no other way to say it. This plane — his new plane — was "Far out!"

❖❖❖

The "JD" on the call sign was not, of course, for Henry John Deutschendorf, Jr., the name given to him by his parents back in Roswell and the name on his pilot's license. Everyone in this little corner of this little airport a few miles from the coast here in Central California knew who was on site and who now owned this Rutan-designed LongEZ experimental aircraft. It was none other than the famed John Denver — songwriter, singer, environmentalist, and lifelong aviation enthusiast. He had searched high and low for a LongEZ of his own and now he had it.

Unlike the other 1200 pilots and craftsmen who had acquired the plans and toiled away for years in garages and hangars across the country and around the world, John Denver never had the time or the inclination to build one of these so-called homebuilt experimental planes on his own. Given his demanding schedule and the money he had in the bank, it made so much more sense to go out and find one in good shape and buy it. And that is what he had done just a few weeks before on September 27, 1997. The previous owner lived in the pastoral western community of Santa Ynez, less than an hour's drive down Highway 101. John had taken a ride in the back seat, just as he had done on a few other rides in the LongEZ, and decided that this was the plane for him. The aircraft had been delivered to the shop in Santa Maria, inspected, cleaned, sanded, re-painted, and flown, and was now ready for its new celebrity owner. John was back in town to pick up the refurbished plane, take his checkout flight, and be on his way back up to Monterey for a game of golf with friends tomorrow morning and then more practice with the plane over picturesque Monterey Bay and Carmel-By-The-Sea in the afternoon. The weather along the coast was great, the people were friendly, and the plane was ready to go.

This little white aircraft was ideal for transporting him across country from one gig to the next, and piloting the LongEZ through the Rockies back home was something to look forward to. The plane was so much more maneuverable and faster than most of the other single-engine alternatives and, with the pusher propeller well behind the cockpit instead of out front where it would obscure the view, so much more enjoyable to fly. Nothing would lie between him and the whole world except the pointed white nose and small forward canard. Flying N555JD over the breathtaking Pacific shoreline or through the alpine passes and valleys in Colorado would be the real-life incarnation of the child's joyous dream — gliding effortlessly from fencetop to fencetop with arms outstretched, propelled magically by thought in any direction through space.

❖❖❖

N555JD LongEZ experimental, the complete and proper name for this specific plane resting on the concrete, was solidly built, well maintained, and capable of everything it was originally designed to do, a plane made for the person consumed with flight and the machines that make it possible. That word "EXPERIMENTAL" stenciled under the pilot's canopy was an important one, however, and meant that many key features of this plane lie outside the rules and restrictions imposed on common production general aviation aircraft. Despite the fact that the LongEZ design was immensely popular and numerous — more numerous, in fact, than many mass-produced production models — the Federal Aviation Administration still considered it beyond the purview of many of its regulations. These "builder-pilots" were the ones putting their lives on the line in their own handmade flying machines, the FAA reasoned; so they were treated with a special set of rules, rules requiring airworthiness and licenses, but rules that allowed considerable design flexibility by the maker and preservation of the fiercely-guarded independence of the so-called "homebuilder."

This particular LongEZ, N555JD, had been hand built in Texas some years before. Like each and every one of the other 1200 LongEZs now flying, this plane was unique in its construction, something to be expected when the same paper plans are executed a thousand times over by hundreds of individuals instead of by a single aircraft factory where consistency in manufacturing and assembly is of paramount importance. It was the builder of this kit aircraft who determined exactly just how much fiberglass and resin to apply to the structure during construction, how much to sand here and there, the amenities and high-end instrumentation in the cockpit, and exactly where to position a display or a control.

The creator of the original design specified every necessary detail in the plans. However, as just about everyone in the aircraft business knew, homebuilders were an independent-minded bunch and often had their own ideas about how certain design details should be carried out. Many homebuilders, including the builder of N555JD, liked to incorporate improvements into the designer's plans, changes that might, at least in their eyes, enhance safety, reduce complexity, improve performance, or just look cool. When finished in Texas in 1987, N555JD LongEZ was structurally sound in the common sense of the term and capable of everything it was required to do. But the plane had a handful of characteristics that made it slightly unusual, especially for a pilot not well rehearsed in the operation of the controls and displays, a pilot who, based on considerable experience with mass-produced aircraft, expected certain things to work in certain ways.

Unlike a creature selected for flight by eons of evolutionary forces, a man and aircraft fly by operating in concert in a system of inputs and outputs between the living and inert, inputs and outputs traveling across the man-machine interface — from the man to the machine and from the machine to the man. It is the interface that makes this system of interworking elements unique, and it is the interface that so often determines the degree to which the system does or does not do what it is intended to do. This beautiful experimental aircraft — this plane that was so capable of everything it was built to do — would soon render one particular and very capable pilot quite incapable of everything that was required of him. This pilot, this lover of flight, this man of song and nature and boundless energy — this man of rhymes — would soon meet his fate for the simplest of reasons.

❖❖❖

As famous as John Denver in his own right, renowned aircraft designer Burt Rutan of Mojave, California conceived the idea and drew the plans for the first LongEZ in the late 1970's. The unconventional approach and overall shape of the plane and its older sister the VariEze sent ripples through the aviation design world and established Rutan as perhaps the most innovative airframe designer around. The concept was just a hint of Rutan's genius in years to come when he and one of his imaginative teams out in the barren desert near Edwards Air Force Base built *Voyager*, the first aircraft of any kind to be flown nonstop and unrefueled around the world. Although considerably larger than the LongEZ, *Voyager* employed many of the features Rutan had incorporated into the VariEze and

LongEZ: a unique shape, composite fiberglass foam structure, low weight, smooth laminar-flow wings, vertical winglets, low drag, canards, and a pusher propeller (and, in the case of *Voyager*, a second propeller in front). *Voyager* ended up hanging from the ceiling in the sacrosanct halls of the Smithsonian National Air and Space Museum in Washington, D.C., and Burt Rutan continued on with his unequaled career designing some of the most innovative flying machines — and spacecraft — ever imagined.

John Denver's particular LongEZ officially came to life on June 12, 1987 when an FAA Airworthiness Inspector from the Houston Flight Standards District Office issued an airworthiness certificate to builder Adrian D. Davis, Jr. Davis had constructed the kit plane using Rutan plans and had submitted the approval application in the amateur-built, experimental category the previous month on May 5. The inspector, after completing the appropriate review of the new plane, checked the box on the application stating "I have found the aircraft described meets the requirements for the certificate requested." The inspector also included a "letter of operating limitations," however, noting that "This aircraft shall contain the placards, listings and instrument markings required by FAR 91.3" (subsequently redesigned 14 CFR 91.9), which stated that key controls and displays were to be clearly identified and marked to aid in their intended operation and that a pilot was to comply with such labels and markings. The inspector had identified a somewhat minor deficiency considering the overall state of completion of this homebuilt, but, for whatever reason, all of the required "placards, listings and instrument markings" were not incorporated into N555JD LongEZ after the airworthiness certificate was issued.

❖❖❖

Back on the tarmac at the Santa Maria Airport, John Denver approached the plane with Roger, his checkout pilot and mechanic. Roger was the same pilot who had taken John for a backseat ride in September and who had transported the plane up to Santa Maria from Santa Ynez for refurbishing. The rocket-like fuselage of the LongEZ was compact and narrow. About two feet back of the pointed front end was the canard, looking something like a 2-by-12 plank inserted through the nose; the elongated bubble cockpit cover began another couple of feet behind that. The pilot's seat was up front and had a high head support that extended almost all the way up to the top of the tinted canopy. Immediately behind the pilot's seat was a bulkhead and behind that a separate passenger compartment. Behind the passenger seat was a floor-to-ceiling firewall and next the engine, a high-output Lycoming O-320-E3D that produced 150 hp. It could push the plane up to 235 mph — making N555JD one of a handful of LongEZs that could attain such speed. Long-swept wings with vertical winglets at the ends and large fuel tanks in the extended wing roots were perhaps the most distinguishing feature of the airframe. Aerodynamic drag reduction was the key to achieving high speed on a compact powerplant, and everything about the shape, especially the overall pusher propeller design, suggested how efficiently it would slip through the air.

There were three landing wheels at the end of spindly mounts, two in the back under the wings and a third in the nose under the canard. The rear landing structures did not retract and were tall enough to ensure that the pusher propeller behind them would not come in contact with the runway during takeoffs and landings. The nose gear was equally tall, but retracted cleanly into the fuselage after takeoff. When fully parked on the ground, the front landing gear could be pulled up into the nose of the LongEZ, which was then dropped down to rest on the pavement, putting the entire plane in an interesting kowtow position.

John and the checkout pilot, both standing next to the plane to make things easier, reviewed the aircraft and its systems, discussing the repainting, the change in the registration number, and basic maintenance. N555JD was nicely decked out for a homebuilt. It had an electrical starter, an electric force bias trim system for both the pitch and roll axis, an electrically actuated speed brake that deployed from the fuselage belly, and a single axis roll autopilot. The control stick containing the electric trim and speed brake switches was located along the right side for right-hand operation.

The two 26-gallon fuel tanks were located on either side of the passenger cockpit, one in each of the wing roots. Each tank was shaped to fit within the unusual form of the wing next to the fuselage. As John had seen during his earlier test ride in the back seat the previous month, the fuel indicators were on the left and right walls of the passenger cockpit, behind the pilot. They were simple sight gauges, each a column of glass filled with fuel and a little floating red ball. With the plane sitting level (or in level flight), the position of the ball displayed the level of the fuel in the tank. But, due to the varying shape of the tank, there was a markedly nonlinear relationship between the height of the floating red ball in each sight gauge and the quantity of fuel in each tank. Pump a few quarts of fuel in each tank — enough to fill the sump and then some — and the ball floated above the bottom of the sight glass; pump in 26 gallons and the little red ball floated all the way up to the top. The fuel sight gauges were not labeled and there were no scale markings, but it all seemed easy enough to understand.

One slightly tricky part was reading the fuel indicators if you were flying without a passenger. Some LongEZ pilots said that they could see the sight gauges from the front pilot's seat by loosening their waist and shoulder straps and craning their heads around to catch a glimpse of the display on each side wall of the rear cockpit. Most just checked the fuel level of each tank while still on the ground rather than trying to read the displays while in the air. Besides, many LongEZs, including N555JD, had a supplemental fuel monitoring system like the *Fuelwatch* device in N555JD that displayed the fuel burn rate and fuel remaining. Use of the device, however, required that it be reset each time fuel was added, something the checkout pilot informed John he had not done due to his lack of familiarity with both the interface and the procedure. Accordingly, John would have to rely on good estimates of the fuel on board when he took off, his flight plan, the floating balls in the sight gauges, and worst-case estimates of fuel consumption until he figured out how to set and use the fuel monitoring instrument.

Each fuel tank was filled from a separate spout on each wing root. The fuel line from each tank was usually routed to a valve in the cockpit — and then back to the engine. As on many aircraft, the pilot selected the tank from which the engine would draw fuel — left or right. Burt Rutan's plans for the LongEZ called for the fuel lines from each tank to be routed up the underside of the fuselage to a valve mounted on the underside of a console directly in front of the pilot, between his legs, and just aft of a small window providing a view down to the landing gear and the earth below. The line out of the valve ran aft, under the console, under the pilot, under the passenger compartment, and back to the engine.

Logically, Burt Rutan's design specified that the little valve handle on the console in front of the pilot be set up so that it was turned to the left to draw fuel from the left wing tank and turned to the right to draw fuel from the right wing tank. The "off" position, in which no fuel would be drawn from either tank, was straight back. Rutan was not a human factors engineer or ergonomist (one who specializes in the design of operator interfaces) but he knew a thing or two about the conditions in which a fuel selector valve might be used and how most pilots would expect it to operate: left is for left and right is for right. It seemed appropriately simple and straightforward.

Rutan was also fully aware of the dangers of having fuel lines running through or under the pilot or passenger compartment. A fire in a cabin, especially one as compact as on the LongEZ, was not something any pilot or passenger wanted to face in the air or on the ground. Accordingly, he designed the underside of the fuselage to withstand a major impact, such as might occur in a crash landing, to protect these lines from damage or rupture. Engineering is often an exercise in trade-offs, and Rutan reasonably balanced the need for an accessible control location with the consequences of running fuel lines near the cockpit. The stronger underside structure would add unwanted weight to the airframe, but provide the necessary margin of safety given the routing of the fuel lines up to the selection valve in front of the pilot.

❖❖❖

The location and operation of the fuel selector valve was one of the first things Roger discussed with John. The Texas builder of N555JD, Adrian D. Davis, Jr., had his own ideas about fuel lines and their routing within the aircraft. This ultimately influenced his decision about where to locate the fuel selector control. Rather than running the fuel lines from the wing up under the cockpit to a selection valve in front of the pilot and then back to the engine, Davis had routed the lines from the left and right tanks directly back to the engine compartment behind the passenger bulkhead in the back of the plane. This way, no fuel lines ran in or under the cockpit. Davis was not aware that designer Burt Rutan had considered this factor and that the original plan with the strengthened underside provided the desirable margin of safety.

But with the fuel tank selector valve now in the engine compartment in the back of the plane, Davis had to devise some way for the pilot to operate the distant valve. The solution seemed straightforward, if not exactly attractive: a long steel and aluminum pole was connected to the valve behind the firewall over the left shoulder of the backseat passenger bulkhead and ran forward 45 inches across the side of the passenger compartment, just above shoulder level, up to the bulkhead behind the pilot's seat. A small handle protruded just within the pilot's reach in the cockpit, behind and over his left shoulder.

The change in orientation of the valve and in its position relative to the pilot meant that the simple control/response relationships envisioned by Rutan no longer applied. There were no labels for the control positions, so Roger showed John Denver how it all worked. To use fuel from the left tank, you had to turn the handle, positioned over and behind you, to the right. Turning the handle to the down position drew fuel from the right tank. The off position was straight up. John thought that it was a little confusing, especially considering that it wasn't easy to even see the control located back behind and over his shoulder. Reaching the handle and flying the plane at the same time was of additional concern. Like the single wrong note in an otherwise perfect melody, it had to be changed; he discussed with Roger his desire to have him reconfigure the fuel lines to bring the selector valve and handle up into position in front of the pilot as specified in the original Rutan design. The shop could certainly make that modification for him, replied Roger, and they discussed a date about a month away when John would drop off the plane for the necessary rework while he was on tour.

There was, however, one more thing that they needed to go over, and it had to do with how John would actually operate the control in flight. The fuel selector valve handle was now over his left shoulder and could be reached and operated only with the right hand. A valve handle forward and center could have been operated by the left or right hand. The trouble with N555JD was that you had to have your right hand on the stick while flying the plane, so operating the fuel selector valve with his right hand was going to be something of a trick. Roger had developed a strategy to deal with this situation, although he had deliberately avoided switching fuel tanks when in the air due to his own dislike of the arrangement. His technique was to loosen the full waist and shoulder straps to give you enough room to wiggle around, engage the autopilot, let go of the control stick with your right hand, twist around almost 90 degrees to the left, and reach for the fuel selector valve handle back over your left shoulder.

Knowing that he would get this whole thing straightened out in a month, but also aware that he was going to be flying LongEZ N555JD until then, John Denver wanted to work out the fuel switching procedure on the ground before going up. Sitting in the cockpit, he mentally rehearsed the steps necessary to operate the controls. He also realized that, due to his moderate stature and the fixed configuration of the cockpit, he had a little trouble depressing the rudder pedals fully. Roger retreated to the hangar and returned in a few minutes with a down-filled seat cushion to place behind John's back. This seemed to help. John was now moved up close enough to the pedals so that he could fully depress them. He was now, however, even a few more inches away from the marginally accessible fuel selector handle back behind his left shoulder.

With various ground preliminaries out of the way and his checkout pilot, Roger, in the back seat, John Denver started the engine, obtained all of his clearances from the tower, taxied to the runway, pointed the nose upwind, and cranked up the throttle. In no time the wheels left the runway and the ground fell away below through the nose gear check window between his outstretched legs. The plane was a dream, sensitive to the touch, slippery, and maneuverable beyond all belief. He gained some altitude and eased into a long sweeping turn, all the while getting the feel of the controls. They turned back towards the airport as if on a large oval racetrack, made a nice long U-turn again, and set up for a little touch-and-go on the runway. With the runway coming up he eased back on the throttle and stick and gently sat the wheels on the deck, then gave it some more gas again and scooted off directly into another takeoff. Roger had him repeat the entire procedure once again, and then together they practiced some slow airspeed maneuvers out away from the airport, all the while communicating over the intercom and headsets that connected John in front to Roger in the back. John appeared to be doing just fine, and Roger had him circle around once again, line up, and bring the plane in for a final landing. It was a great first flight, if only a little short.

❖❖❖

His checkout flight completed, John packed his gear in N555JD and said his thank-yous and goodbyes to Roger and the folks at the Santa Maria airport. Back out at the plane, Roger told him that he had something on the order of 15 or more gallons of fuel on board, which was about twice as much as he should expect to consume in the hour-long flight up to Monterey. About one-third of the fuel was in the left tank and two-thirds in the right.

Sitting in the cockpit one last time before departing, John again asked Roger to go over the unmarked positions of the fuel selector handle to make certain that he had it right. He had to remember that right was for the left tank, down was for the right tank, and up was for off. OK, he could remember that. And if by chance he should accidentally run a tank dry, the engine would "pop" loudly as it lost compression and quit. He didn't want to be changing tanks under such circumstances, but it was good to know that it could be done if by chance he used all of the fuel in one tank. With the fuel selector moved downward to draw fuel from the fuller right tank, John Denver obtained his clearances from the tower and departed northward to Monterey.

❖❖❖

It was always a difficult thing to explain to a non-pilot. There was something about mastering all of the procedures, navigation rules, instrumentation, and skills involved in flight that John found deeply satisfying and even relaxing. Perhaps it was that piloting was so different from songwriting. The latter required hours of stumbling through possibilities and the occasional moment of pure inspiration when lyrics and melody worked perfectly together; the former demanded discipline, order, and precise execution. It was also possible, of course, that his enjoyment of flight was due to the simple fact that he was raised on various Air Force bases in the 1950's, surrounded by all manner of powerful and exotic flying machines, or that it was often his father who was up in the sky in the pilot's seat as the loud jet roared overhead. At any rate, it didn't really matter. Flying was fun, he could certainly afford it, and piloting N555JD was a blast.

Many years before this day, John had reached and surpassed the level of skill one might call an expert. He was not one of those weekend flyers who spent a nervous Sunday morning looping around the same pattern in a Cessna 172. John Denver was a highly experienced pilot, having logged more than 2,750 hours of stick time in everything from a glider, to a seaplane, to a Learjet. He had owned many planes and had once even flown a jet fighter — courtesy of his father and the Air Force.

LongEZ N555JD actually felt a little bit like that jet fighter, only smaller and so much lighter. It was certainly nimble, and it took only the slightest movement of the hand and foot controls to bank, climb, or dive. Unlike a conventional airplane where the rudder pedals were connected to a single vertical fin at the back of the plane, the LongEZ pedals controlled the two vertical rudders on the tips of the left and right wings. Pressing the right rudder pedal moved only the right rudder in an outboard direction, increasing drag and turning the plane with a right yaw. The two rudders were very effective at turning the plane, and, because they were tall and above the longitudinal center of gravity of the entire aircraft, rudder activation also produced a pitch-up moment along with the yaw. This required John to make a little downward pitch input on the stick in his right hand to keep the nose generally level whenever he turned using the pedals. Coordination was all part of flying through three dimensions, and it was great getting the feel of a new machine.

After his climb out away from the airport and once up to cruising altitude, John settled back into the loose seat cushion behind his back. It was nice of Roger to have found it for him in the shop. The cushion wasn't too uncomfortable, but again this was something that he would probably get changed at some point in the near future. This was, after all, now *his* airplane and he should set it up to his liking.

Off to the right down below was Highway 101. To the left was the productive farm land of the Santa Maria Valley and, beyond, the coastline and the Pacific. The little beach towns of Oceano, Avila, and Cayucos came and went below, as did San Simeon and serpentine Highway 1 along the Big Sur coast.

Within no time he was approaching the Carmel Valley up ahead. He had covered a lot of ground in under an hour. John Denver contacted Monterey airport, obtained landing clearance, and brought the LongEZ N555JD in for a smooth touchdown. He taxied to a hangar where he had made arrangements to park the plane indoors for the night. It had been a great day. Now it was time for some relaxation with friends, a good night's sleep, and a game of golf tomorrow morning followed by more practice with his new aircraft late in the afternoon.

❖❖❖

After a leisurely breakfast, John Denver enjoyed the scheduled game of golf on a spectacularly beautiful course and afterward moved on to a late lunch. There, he declined all offers of drinks from his friends, knowing the importance — and illegality — of such things given his plans to fly later that afternoon.

Back in the hangar where he had parked his LongEZ at the airport, he met up with a maintenance technician who offered a helping hand. They discussed the plane, as pilots like to do, and talked about the cockpit and controls. John mentioned the situation with the fuel selector valve handle, and the technician retrieved a set of vise grip pliers that they clamped onto the valve handle over the pilot's left shoulder to make the control a little more accessible and easier to turn. It seemed a bit of a kludge, however, and John decided that he would stick with the system as it was, rather than messing around with the vise grips. Should he need to change fuel tanks, which would be unlikely on his short flight, he would just turn on the autopilot, as he had discussed with the checkout pilot in Santa Maria, release the stick, and reach up and turn the valve over his shoulder with his right hand.

With the assistance of the congenial technician, John started working through his preflight activities. The flight up from Santa Maria the day before had consumed some portion of the roughly 15 gallons of fuel with which he had started out; the question, of course, was how much. He asked the maintenance technician about the fuel quantity in the right and left tanks. The mechanic looked over the open rear cockpit to inspect the sight gauges in the back seat. The red ball in the right tube was floating just below the halfway mark; the red ball in the left tube was floating just shy of a quarter of the way from the bottom. You have "less than half in the right tank and less than a quarter in the left tank." "Do you want to add fuel?" he asked.

John thought for a moment about what the technician had said, his flight up from Santa Maria, the two 26-gallon tanks, and his plans for the afternoon. With nearly half in the right tank and almost a quarter in the left, he reasoned, he should have plenty of fuel. After all, he was only going to do a few touch-and-goes and get out over the ocean for a little while. "No thanks," he said. He was going to be up for no more than one hour and did not need to take on fuel.

The interchange about fuel levels did, however, raise the recurring point about seeing the fuel sight gauges from the pilot's seat. The maintenance technician had another idea; he once again retreated for a minute and returned with a small shop mirror mounted on a long thin handle used to look around corners in tight spaces. John could actually see the two fuel indicators back behind him on the left and right walls of the passenger compartment while holding the mirror, but again it certainly wasn't optimal. He thanked the mechanic and stowed the mirror in the cockpit.

For the next twenty minutes, John Denver stepped through his preflight procedures with occasional assistance from the maintenance technician. In addition to various routine inspections, John wanted to check for contaminants in the fuel, so he borrowed a fuel sump cup and drained a small amount of fuel from each tank's sump. The fuel was clear and everything looked as it was supposed to.

With the plane out of the hangar and on the tarmac, John straightened up the feather-filled seat cushion, climbed up into the cockpit, sat down, and strapped himself in. The maintenance technician wished John a good flight and gathered up his tools as John continued with his preflight activities from the cockpit. He checked the operation of the control surfaces, verified control positions, put on his headset, looked around to see that everything was clear, and started the engine. It roared to life for a few seconds, but soon quit. John twisted around to the left and reached up to the fuel selector valve handle. It was pointed up, the "off position." There must have been just enough fuel in the line to start the engine. He turned it to the right position to draw fuel from the left tank and turned back around, straight in his seat. John saw the maintenance technician, who had obviously heard the engine start and then die, walking back out from the hangar to see what was going on. He waved to him and gave him a thumb's up signal, indicating that everything was fine. John then looked around again to make sure everything was clear and restarted the engine. It fired right up and continued to run smoothly. The maintenance technician returned the wave and walked back into the hangar.

John contacted ground control at the Monterey Peninsula Airport Air Traffic Control Tower and obtained a taxi-for-takeoff clearance from out in front of the hangar. It was 5:02 in the afternoon on October 12, 1997. He taxied toward the east end of the runway for a planned takeoff west out over the ocean. At 5:09 he contacted the local controller, reported that he was ready for takeoff on runway 28, and requested permission to stay in the traffic pattern for some touch-and-go landings, circling around a large imaginary race track to practice takeoffs, flying, and landings. Three minutes later at 5:12 the controller informed him that he was cleared for takeoff. He straightened out the nose of LongEZ N555JD, increased the throttle, and let up on the brakes. The plane accelerated smoothly off the line and quickly left the ground. The west end of the runway fell away through the nose gear window, as did the Navy Golf Course and the Fairgrounds and Highway 1. Seconds later he was above the beach and Monterey Bay, flying over Cannery Row and the Aquarium. John executed a nicely rounded U-turn and headed east to make another U-turn so that he could line up headed west again, drop in, and make a touch-and-go on the same spot where he had taken off just minutes before. He eased back on the throttle, gently brought the plane down on the runway, accelerated back up to speed and took off again over the Pacific. He looped around again for a second time in the traffic pattern, made another touch-and-go, rose above the bay and Point Pinos and prepared to loop back again for a final touch-and-go.

The plane indeed flew like a dream. All he had to do was think about where he wanted to go and it took him there. The tiniest movement of his hand, the smallest pressure by his feet — so small that it was easy to forget that you had to move anything at all — broadcast his thoughts to the plane. No, that wasn't right. It was really more like there was no plane at all. He was flying through space exactly as he had dreamed it as a child, pushing away from the top of the fence and gliding over the backyard, his arms outstretched and his heart full of joy. This is what it was all about.

❖❖❖

During the preceding 15 minutes of touch-and-goes the little red ball on the left wall of the passenger compartment behind

John had fallen precipitously toward the bottom of the vertical tube as the engine drew down the fuel in the left tank. Neither John Denver nor the maintenance technician on the ground knew that the *level* of fuel in the tank and the corresponding position of the red ball in the sight glass did not directly match the actual *volume* of fuel remaining. When he left the hangar only 25 minutes before, the 26-gallon tank was not (as suggested by the quarter-way position of the red float in the unmarked, unscaled, and transparent tube) nearly one-fourth full with over 6 gallons of fuel. It was closer to 3 gallons. And the right tank, rather than having just under half of 26 gallons as inferred by the floating red ball in the unmarked glass tube, contained about 6 gallons. Although built to specifications, each tank was not the same size from top to bottom. The lower quarter of each tank — at least as measured by its height — held much less than the middle or upper quarters. Not only was each sight glass unlabeled, it presented a decidedly nonlinear representation of the fuel volume. In actuality, there was now not even enough fuel in the left tank to fill the one-quart sump. LongEZ was in serious danger of running out of gas in the left tank at the most inopportune moment, and John Denver would be faced with the problem of operating the confusing and inaccessible fuel selector valve to switch to the right tank.

At 5:27 John communicated with the controller for the seventh time in the past dozen minutes, informing the tower that he was going to leave the traffic pattern after one more touch-and-go and fly out over the ocean for a few minutes. The controller acknowledged his transmission and requested him to recycle his transponder code, which he did. With permission for his approach, he made the final adjustments to his westward turn and tilted the white nose of N555JD slightly downward toward the runway. He eased back on the throttle, dropped the plane down gently, kissed the earth with the wheels one last time, pushed up the throttle and rose up and away over the Pacific and the setting western sun. This time he hit the landing gear control, and the nose gear retracted smoothly into the fuselage.

Moments later, before passing through 500 feet, just beyond Cannery Row and the Monterey Bay Aquarium, the engine sputtered and then "popped" as it starved for fuel and lost compression. A dozen residents and tourists enjoying the October sunset raised their faces to the odd sound from the strange white craft moving across the blue sky not so very high overhead.

At that moment John Denver knew, no doubt, exactly what had happened. There wasn't time to loosen the straps or set the autopilot. He turned frenziedly in his restraints, his left side against the feather-filled cushion behind him, and reached hard with his right hand for the fuel tank selector valve well behind and over his left shoulder. In so doing he applied just the slightest bit of pressure on the right pedal. The tall vertical rudder at the end of the long right wing moved outward, increasing drag and turning the plane to the right. Simultaneously, the nose pitched upward. Without his hand on the stick to counteract the natural action of the plane, the nose continued to rise up, and within a half second LongEZ N555JD had rolled hard right and over with her nose aimed straight down to the sea. The plane drilled nose-first through the surface just off Lover's Point, shattering into a mass of fiberglass, metal, flesh, and bone as it slammed into the rocks on the bottom 30 feet below the surface.

REFERENCES AND NOTES

Death of a troubadour (1997). *Houston Chronicle*, October 13.

John Denver confirmed dead (1997). *Cable News Network www site,* October 13.

Denver's death seen as loss for Hawaii environmental movement (1997). *Associated Press www site,* October 14.

Denver sober when plane crashed (1997). *The Seattle Times www site,* October 29.

Garrison, P. (1999). Aftermath – John Denver. *Flying Magazine,* May.

John Denver crash report called flawed (1999). *Denver Post,* January 27.

John Denver update (1997). *Sun Tsu's Newswire www site #PIRN 9776.*

John Denver update – no gas? (1997). *Sun Tsu's Newswire www site #PIRN 7880.*

Kreidel, R. (2000). *Personal communication* (LongEZ pilot and accident consultant).

NTSB: Denver's plane short on fuel (1999). *Associated Press www site,* January 26.

NTSB determines John Denver's crash caused by poor placement of fuel selector handle diverting his attention during flight [press release] (1999). Washington, D.C.: National Transportation Safety Board, January 26.

NTSB Investigation Report LAX 98FA008: Report on crash of Adrian

Davis LongEZ, N555JD, piloted by John Denver on October 12, 1997 at Pacific Grove, CA. (1999). Washington, D.C.: National Transportation Safety Board.

Plane design faulted in Denver crash (1999). *Washington Post*, January 27.

Three factors cited in probe of singer's plane crash (1998). *Santa Barbara News Press*, June 23.

Sec. 91.9 civil aircraft flight manual. Markings and placard requirements, 14 CFR 91.9 [formerly designated FAR 91.3] (1990). Washington, D.C.: Federal Aviation Administration, August 18 .

A KID IN A CAR

It was afternoon and time for a nap. Her father knew it, Zoie wanted it in truth, but she was not about to admit it with that touch of obstinacy or falling eyelids and nodding head of a two-year-old who has been up for most of the day. Life was too full, too real, too fun, too happy to miss even an hour. Nevertheless, he knew what was right and picked her up with all the care and love of a parent, and she held on willingly to her father's side under his strong arm as he opened the door of his pickup truck with his other hand and slid the front seat forward and leaned in and placed her gently in the back. The truck was his extended-cab Ford, the one with the power windows and all of the options and comfortable bench seat in back with more than enough room for a toddler to lie down and take a nap on a pleasant day in Anthony, Kansas while her father worked outside for a short while just a few steps away.

It was nice inside the truck. The grown-up talk and small noises from outside faded into the distance and the padded bench seat under her head and back felt like her bed at home when the sheets were tight and newly made. The ceiling above was a soft fuzzy carpet, and a light breeze flowed through the comfortable cab from one side to the other. Unhurried white clouds slid across the upside-down picture framed by the open window just behind and above her head. They were pretty white clouds with funny heads and big puffy bodies with little tails that seemed to wiggle as they drifted across the blue picture from one side to the other. It was fun to watch the clouds.

❖❖❖

Then from outside, amid the little noises and the occasional clank of a tool and hushed voices of grownups who knew that someone had been put down for a nap, there was a sound — the sound of something alive, the sound of quick light steps and paws on pavement, the unmistakable musical whimper of a real dog, not an imaginary dog in the sky. It sounded like a friendly dog, a big puppy looking for attention and a pat on the head, and she most certainly knew that its tail was wagging.

Zoie rolled over onto her side and then onto her hands and knees and crawled across the seat to the open window. She probably grabbed the arm rest first, then reached up the door to hold the ledge where the window was retracted within the wall, and pulled herself up. Her head popped above the window sill and she looked toward the sound. It *was* a dog! A fluffy happy dog, and its tail was wagging back and forth a million times a minute! She raised one knee and then the other up onto the armrest to prop herself up to look out the open window at the real dog just outside. Her face was now just beyond the plane of the retracted glass and her knees on the armrest bore her weight.

The dog saw Zoie framed in the window and turned toward her. Her head came out a little more in anticipation and her knees shifted across the armrest, an armrest with a rocker switch on its upper surface — the control for the electric power window. Pressing one side of the switch made the glass retract and go down, which, of course, it already was. But pressing the other side made the glass go up. Like the radio and the other electric-powered parts in the cab, the power windows worked only when the key had been left in the ignition and turned to the "accessory" position, which, tragically, was the case on this otherwise pleasant day in Kansas.

❖❖❖

Her knee came to rest on the switch, the electric motor started to spin, and the window launched up and out of its slot; it was a very strong window made of shatterproof safety glass with a dull but solid edge. It caught her in the delicate intersection between her chin and neck and took her up to the top until the back of her neck below the skull pressed hard against the upper edge of the steel frame and the full weight of her dangling body bore down on the edge of the sheet of glass pressed against her two-year-old esophagus. She did not make a sound.

Roughly one minute passed before the two-year-old was seen. The door was opened as fast as humanly possible and the window lowered and her lifeless body pulled from the truck.

❖❖❖

Zoie Gate's was but one of 42 strangulations from power windows identified by an informal survey of newspaper articles covering a number of decades conducted by the nonprofit group Kids 'N Cars in 2002. Each of these accidents involved a disturbingly similar set of circumstances: an unattended toddler, a key left in the ignition, a traditional rocker-type switch design, and a Ford, General Motors, or Chrysler product. The Kids 'N Cars research identified no cases involving recessed lever-type power window switches which were designed to avoid inadvertent actuation, especially by a child. Such switches are found on luxury vehicles and foreign cars designed to a different set of standards. To date, the "big three" auto manufacturers have maintained that their designs are safe, that keys should never be left in the ignition, and that children should never be left unattended in cars, which of course is true. Be that as it may, the paucity of such accidents in cars having recessed power window switches speaks for itself.

Furthermore, although existing Federal standards limit the maximum force of a power window to reduce the likelihood of finger amputations and the like, this force — and its concentration on the small surface area of the edge of a plate of glass — easily exceeds that which is sufficient to strangle a toddler. The auto-reversing power window — one which retracts when it encounters an obstruction — is available on some high-end vehicles and cars having an "express-up" function in which the control need not be held while the window goes up. Also, rear window lockout switches are available on many vehicles, but frequently are not used.

REFERENCES AND NOTES

During the last five years, over 235 young children have died as a result of being left in unattended cars [press release] (2002). San Francisco, CA: KIDS 'N CARS, February 10.

Fact sheet: injuries/deaths of children left unattended in or around motor vehicles (2004). U.S. Department of Health and Human Services, Centers for Disease Control, National Center for Injury Control and Prevention, July 5.

Laboratory test procedure for FMVSS 118: power-operated window, partition, and roof panel systems, TP-118-03 (1994). Washington, D.C.: Department of Transportation, Office of Vehicle Safety Compliance.

O'Donnel, J. (2992). Power car windows can strangle. *USA Today,*

May 22.

Research Note: injuries associated with hazards involving motor vehicle power windows (1997). Washington, D.C.: Department of Transportation, National Highway Traffic Safety Administration, May.

Wright, J. (2003). Power window safety at issue. *Los Angeles Times*, January 15, G1.

Wright, J. (2003). Making power windows safer. *Los Angeles Times*, October 1, G1.

THE PERILOUS PLUNGE

Fried chicken, mashed potatoes, chicken gravy, buttermilk biscuits, sweetened rhubarb, and the piece of boysenberry pie *a la mode*. What more could a hungry soul ask for? On Friday night, September 21, 2001 the offerings available to Lori Mason-Larez and the other guests at Knott's Chicken Dinner Restaurant in Buena Park, California were much as they had been for 67 years. There was just a lot more of it nowadays and an astoundingly greater number of hungry customers and generous meals served from the same location off Beach Boulevard every day except Christmas, 364 days per year, an average of 3,021 meals per day, in excess of one million meals each and every year.

The menu was basically the same as in 1934, but otherwise Knott's was vastly different than when Walter and Cordelia Knott first opened Knott's Berry Place out in the center of rural Orange County in Southern California. It started as a berry farm with a roadside stand to hock Walter's fresh fruit and Cordelia's tasty preserves to passing motorists. Walter popularized the boysenberry, named after his neighbor, Mr. Boysen. Business picked up quickly when Cordelia began selling the chicken dinners with help from their three daughters, transforming the place from a roadside fruit stand and tea room into a bustling eating establishment, all in the middle of the Great Depression. Customers were soon lined up outside the door.

Within a few profitable years, Walter saw that the folks

needed to be entertained while they waited for a table at the newly named Chicken House Restaurant at Knott's Berry Farm. He started buying up old western ghost towns and moving interesting buildings and rusty mining gear to a transplanted ghost town of his own design out behind the restaurant. In time there was a shooting gallery, a real stagecoach with a team of horses, a mule ride with actual mules, and even an antique narrow-gage steam engine and train that ran on its own track. The amusements eventually drew more visitors than the Chicken House Restaurant, and by 1968 they were forced to fence in the whole operation and charge a small admission to the park. You could still pan for gold, watch the flour mill turn, see the girl spin wool on an old spinning wheel, and even find a peacock feather or two, but it would cost you a couple of dollars — a small price to pay for a few hours of pleasant and safe entertainment.

But times change and people change. Walter and Cordelia, and eventually their daughters, passed on after long and active lives, and competing theme parks with more thrilling amusements delivering instant gratification of rapid acceleration and centrifugal force, not fantasy reenactments of the old West, popped up around the state. By the last decade of the century, the Knott heirs knew that if Knott's was to survive they had to take a plunge of their own and they eventually sold out. The profitable line of packaged foods became part of ConAgra, and in 2001 Knott's Berry Farm was known as *Knott's Theme Park,* "America's favorite theme park" with "world class rides" and "high-octane thrills," run by Cedar Fair of Sandusky, Ohio, owner and operator of a half-dozen big-time amusement parks spread around the United States. The most popular attractions now were the new rides built out behind the old Ghost Town, especially the rides that gave a minute or two of visceral thrills, like the *Supreme Scream, Grand Slammer,* and, especially, the

Perilous Plunge.

❖❖❖

At a quarter past 10 o'clock on Friday night, September 21, 2001, a few hours after dining at the Chicken Dinner Restaurant, Mrs. Lori Mason-Larez of Duarte and three of her five children stood in the crowded line anxiously awaiting their turn to board the 24-person boat for a spin on the *Perilous Plunge* amusement ride billed as "the world's tallest and steepest watercoaster." Another of her children, 9-year-old Marty John, was going to watch and not ride; he was out with his aunt, Shirley Roman, at a good vantage point to see the faces of his mother, brother, and sisters when their open boat tipped over the edge of the *Perilous Plunge* way, way up in the air, practically falling down the unbelievably steep slope 15 degrees off of vertical from the dark sky high overhead. The boat could not actually come out of the trough when it shot down the watery slope because it was mechanically attached. Regardless, the whole thing looked rather perilous to anyone standing in line on the boarding ramp waiting to get on.

Once Lori, the three kids, and other riders climbed into the boat it would take only a minute until it was released into the flume and made its way around to the base of the steeply angled trough headed up. There the boat would latch onto the chain drive and be pulled up out of the water and high into the sky, just like on a roller coaster. It would make a little half-turn while floating in the elevated water trough 120 feet in the air and then take the plunge over the edge. They would reach a speed of over 50 miles per hour on the way down, then decelerate as the bow hit the water at the bottom, pushing a two-story wave into the air. Based on the deafening screams heard every minute when a boatload of riders went over the "falls," it would be quite a show when the Mason-Larez clan took the plunge. Everyone would, of course, get thoroughly drenched — but that was part of the fun. The ride was said to simulate going over Niagara Falls, and, given its height of nearly 120 feet, this was not too far off the mark. It would be over quickly, more quickly, as things would turn out, for one rider in particular.

❖❖❖

Shirley Roman, Lori's sister, worked for the Target department store chain and had acquired the tickets for Knott's at a special discount arranged by Knott's and Target. Shirley brought her own daughter, her sister Lori, and four of Lori's five kids as guests. The husbands had opted to stay home. Earlier that evening, after dining at the Chicken Dinner Restaurant, the group had walked through what was left of Ghost Town and gone on an assortment of rides. Lori found the courage to ride *Jaguar,* a reasonably tame roller coaster with not as many turns and upside-down maneuvers, but she opted to sit out when the group went on *Montezooma's Revenge,* a serious roller coaster with two upright loops. Taking the kids to a theme park was one thing, but going on a thrill ride that looked and sounded as if it might be a bit "over the top" in terms of personal discomfort was another.

As seen from above, the course of the *Perilous Plunge's* trough that carried the water and boat was a big figure eight. The bottom loop circled over a small pond of water, and this is where the boat slowed to a stop so that riders could board from the platform on which Lori and her kids now waited in line. The other half of the figure eight was 120 feet in the air, and the cross point of the "up" trough with its integrated chain lift and the "down" trough where the boat zoomed back down was about halfway up. It was a brief ride as amusement park rides go. The boat made only one complete circuit of the figure eight: the loop around the bottom, the steep rise up to the top, a turnaround, and then over the edge and down the steep slide with the big splash at the bottom where the boat came to rest and the riders, drenched from head to toe, disembarked in laughter and disbelief.

❖❖❖

Judging by the screams of the riders a half-minute before, it sounded as if the whole experience would be more terrifying than thrilling — especially the moment when the boat tipped over the edge more than ten stories high and started its descent down at 75 degrees. But the riders sitting in the boat that now drifted up to the loading dock were all laughing and chatting away, so it couldn't be all that unbearable. There was the sign at the beginning of the line that said that people with heart problems, back or neck trouble, or those who had recently had surgery should not go on the ride. Neither Lori nor the kids had any of these problems, so there was no reason to think that the *Perilous Plunge* might be dangerous for any member of the family.

❖❖❖

Built at a cost of 9 million dollars and in operation for about a year, the *Perilous Plunge* was designed and constructed by Intamin AG of Switzerland, one of the largest manufacturers of amusement park thrill rides in the world. Passengers in the boat obviously had to be restrained, especially for the plunge over the edge and the long ride down, so Intamin incorporated a common T-shaped restraining bar for each seat. The seated passengers would pull the bar toward them, straddling the lower vertical part of the T-bar between the legs, and pull the horizontal part down on top of the upper thighs. The bar had a ratcheting effect and could be pulled down far to sit tight on the legs of a small or thin person, or pulled down only partially if the rider was large. Once engaged and checked by the ride attendant at the loading dock, the bar stayed in place until the boat completed the figure-eight circuit of the ride and returned to the loading dock. There was also a horizontal grab handle up at about chest height in front of each seat.

Cedar Fair, owner of Knott's, felt that the T-bar by itself was insufficient for restraining passengers and asked Intamin to add a seat belt to each of the 24 seats in each boat. Intamin made it clear that the seat belts were unnecessary for the safety of riders, but wanted to appease their client and went ahead and added the seat belts. People come in all shapes and sizes, of course, and the seat belt was made long enough to encircle a girth in excess of 60 inches.

❖❖❖

The line was now short and they would be among those boarding the next boat, Boat No. 1. It had six rows of high-backed red plastic seats, and each row held four passengers. Boat No. 1 had just come down the flume, making a truly impressive wave as it bottomed out at the foot of the steep slide. It drifted up to the loading platform, its occupants wet but happy. The late summer evening was pleasant and cool, just the right temperature for wearing a light sweater. The soaked riders now unbuckling their seat belts left no doubt that Lori's blue sweater was going to be dripping wet when this was all over.

The single broad line of people split up into six single-file lines — one for each of the six rows of seats on Boat No. 1. Lori and the kids had been assigned to the fourth row. The kids were excited. The time had come. They stepped off the dock and over the side and down in. It was a bit wet inside and the boat moved slightly as the passengers stepped around and scooted across. Lori took seat 13 and her daughter Darlene sat to her right. The seat belts were somewhere down between the seats, and Lori slid her hands down beside her thighs to find the belts that belonged to her. Darlene and the other kids were quick to catch on and had already found and buckled their own seat belts, each one sliding the loose end through the buckle and tightening it low and tight across the waist. Darlene, being the helpful daughter that she was, turned to the side and helped her mother find the two pieces of her seat belt, threading the parts together, and tightening it down. Lori, like every other passenger, grabbed the top part of the T-bar up by her knees and pulled it back toward her. The ride attendants looked everything over. Everyone was buckled in and the restraint bars were in place. A few words were said and the boat rolled off of its restraints and floated out into the current of the water-filled trough with a boatload of giddy thrill seekers at Knott's Theme Park. The time was 10:19 p.m.

❖❖❖

They floated softly in the current. It was comfortable, and the sounds of the park faded into the void of the dark sky overhead. The boat drifted around its half-circle course. In under a minute they were facing the tall trough to take them up, up high into the sky above the crowd. The mechanisms took hold of the underside of the boat with a clank and pulled it forward and then upward. There was no going back now. The bow rose, and in a moment they were angled steeply and being towed almost straight up, it seemed, into the sky, higher and higher, until they were 10 stories, then 11 stories, then a full 12 stories up above the rest of Knott's Theme Park and the bustling crowds below.

Then the bow flipped over back down level as the boat crested the vertical trough onto the horizontal track. It smacked flat off the drive mechanism of the lift and into the water in the elevated trough. They were floating again, floating in a boat in a water-filled trough 120 feet up in the air at 10:20 at night above Southern California. There was Walter Knott's old Ghost Town down below and off in the distance Beach Boulevard and the lights of Orange County. The lights seemed to spread out forever.

By now the seat belt that had originally been strapped around Lori Mason-Larez's 58-inch stomach had worked its way down to a position near her pelvis and legs. At 5 feet 8 inches and 292 pounds, Lori Mason-Larez was a very large woman whose center of gravity was considerably higher than that of a woman of average weight. The T-shaped restraining bar was positioned exactly where it was when the boat left the dock, but, due to her girth, the horizontal part could not be pressed firmly against the muscles of the upper leg and instead was up against the soft fluid tissue around her abdomen. Lori turned and exchanged some friendly but nervous words with the people in the row of seats behind her, then looked over at the kids. They started to scream. The drop-off ahead was as abrupt as the edge of a cliff. The sound of the water flowing over the edge grew loud. She held on to the grab bar in front.

Boat No. 1 dove over the precipice with a force strong enough to fling out anything or anyone not firmly restrained. The bow nosed down, the stern flipped up. Lori Mason-Larez, her hands gripping the handle in front of her with all her might, unable to overcome the substantial force of her own accelerated mass, popped out of the boat up high in the air like a blue champagne cork. Daughter Darlene's thrill ride scream turned into one of utter terror as the blur of a large blue body flew out and down past the boat within the heavy spray of water coming off the downshoot at the top of the flume, impacting the rail of the trough ahead and below them, then the support girders, and on down to the shallow pool 120 feet below. The boat accelerated to 50 miles per hour, then bottomed out at the base of the flume and pushed out a spectacular 40-foot spray of water. The waves died down, the water slowed, and Boat No. 1 drifted into the dock in the current. Seat No. 13 was empty. Its lap bar was in its original position and the seat belt was latched, lying flat on the seat bottom. Lori Mason-Larez was declared dead one hour later at the West Anaheim Medical Center.

❖❖❖

Knott's immediately closed down the *Perilous Plunge*, the investigators were called in, and the attorneys sharpened their pencils. The first report was from the Orange County Coroner who revealed that Lori Mason-Larez had suffered a fractured skull, a fractured arm, and a severed leg during her horrific fall. The report also disclosed her unusual dimensions and weight. Not unlike other amusement park patrons seriously injured or killed in recent years, she was at one of two extreme ends of the distribution of body size, with physical attributes not well accommodated by the particular restraint system. By October, the family had filed a wrongful death suit seeking 30 million dollars in damages from Knott's parent company, Cedar Fair, and the manufacturer of the *Perilous Plunge*, Intamin AG.

Investigators from the California Division of Occupational Safety and Health issued their report five months later and identified both the cause of the accident and a dozen required changes to the ride. With her 50-inch hip circumference, 58-inch abdomen, 5 feet 8 inch stature, and weight of 292 pounds, the rider could not be adequately restrained by the seat belt and the T-bar during the rapid teeter-totter motion of the boat as it topped over the edge headed down the shoot. Her high center of gravity and low strength-to-mass ratio further limited her ability to hold herself down in her seat. Given these facts, her ejection from the boat was a near certainty. And although the ride's operating manual produced by Intamin AG stated that any person who could not fit within a seat and the restraint system should not be allowed to ride, the procedures were loosely interpreted and enforced by the ride's staff so as not to offend certain patrons.

On June 1, 2002 Knott's General Manager Jack Falfas re-opened the *Perilous Plunge* with a test ride before the media and park visitors. Among other things, new signs listing maximum height and weight limits had been posted, a mock-up of the seats and restraint systems for examination by patrons had been placed in the waiting line, and a new set of verbal instructions for passengers had been adopted. Most importantly, and against the advice of Intamin AG, Knott's added a four-point, double over-the-shoulder restraint system to each seat. The family's lawsuit against Cedar Fair and Intamin AG was settled on March 28, 2003 for an undisclosed amount.

REFERENCES AND NOTES

Basheda, L. (2002). Perilous Plunge reopens 8 months after fatality. Amusement. The thrill ride resumes at Knott's after closing in September when a woman fell off. *The Orange County Register*, June 2.

Deaths prompt industry investigation of theme park rides (2001). *Associated Press*, November 4.

Herubin, D. (2002). Ride restraints not designed for the obese, maker says. *The Orange County Register,* March 20.

Herubin, D., Luna, N., and Saavedra, T. (2001). Ride-fall report cites obesity. The manufacturer of the Perilous Plunge says the woman who died was too big for the ride's safety restraint. *The Orange County Register,* October 23.

Himmelberg, M. (2002). Weighty warning from Knott's. Safety. Death of visitor prompts park to post signs stating guests of 'extreme size' can't ride certain coasters. *The Orange County Register,* April 6.

Himmelberg, M. (2002). Knott's to add restraints to coaster. Theme parks. The addition of harnesses to Perilous Plunge comes after fatality last year. But ride maker argues move. *The Orange County Register,* May 10.

Himmelberg, M. (2002). Knott's water coaster returns. Safety. After 8-month closure following fatality, state OKs modified Perilous Plunge. *The Orange County Register,* June 2002.

Himmelberg, M. (2003). Deal reached in fatal Knott's ride. *The Orange County Register,* March 29.

Himmelberg, M., Saavedra, T., and Herubin, D. (2002). Restraints, obesity cited in ride death. Report experts say those factors send a warning to the theme-park industry. *The Orange County Register,* March 20.

Keith, L. D. (2002). Amusement park death investigated. *Associated Press,* March 19.

Luna, N. and Hardesty, G. (2001). Family sues Knott's over death on ride. Woman's relatives say park, manufacturer were negligent in operation of Perilous Plunge. *The Orange County Register*, October 10.

Quach, H. K. (2002). Board grapples with rules for theme-park warnings. Safety: officials, businesses debate best way to protect extremely small or large passengers. *The Orange County Register*, June 21.

Saavedra, T., Herùbin, D., and Knap, C. (2001). Register investigations. Ride safety not equal. Parks: developers of restraint systems are challenged to design safety devices for people of various sizes. *The Orange County Register*, November 4.

State: restraints lacking on Knott's ride where rider died (2002). *The North County Times*, March 20.

Theme park tour: Chicken Dinner Restaurant overview (2003). *Knott's Southern California Resort www site.*

Theme park tour: Perilous Plunge (2003). *Knott's Southern California Resort www site.*

Yoshino, K. (2001). Weight may have been a factor in death at Knotts. *Los Angeles Times*, September 25.

TITANIC'S WAKE

"Hey Walter! Come over here. You've got to see this," called out David Durand to his friend and coworker.

The scene below on the Chicago River was enough to send a shiver up anyone's spine, and the vantage point from the third story of the Watson Warehouse on the north bank of the river that ran east and west through the middle of downtown Chicago was disturbing, but also oddly comical. The time was 7:15 in the morning, Saturday, July 24, 1915. The stage was set, the players had arrived, the curtain had gone up. And like anxious theater patrons in the front row of the balcony, the two warehouse employees had an up-close and personal view of the drama's final act. David Durand's hunch that all was not right was right on the mark. The two were about to witness theatrical tragedy: a compelling plot, a tension-filled setting, a cast of thousands, a calamity of unparalleled and unthinkable consequence. The decisions of designers, a profit motive, the legacy of the *Titanic,* the ill-conceived directives of politicians, and the behavior of an energetic crowd of young Americans were about to converge in one horrible moment in time down on the river near the Clark Street Bridge in downtown Chicago.[3]

Walter Perry, now with David Durand at the window, took a few seconds to comprehend the sight on the river. His response said it all: "My God!"

[3] This story is based principally on the definitive book on the subject, Eastland: Legacy of the Titanic, by G. W. Hilton.

❖❖❖

Tied up to the dock on the south bank of the well-developed Chicago River at the Chicago and South Haven wharf between Clark and La Salle, diagonally across from the third-story window in the Watson Warehouse, sat the biggest and most glamorous of the Great Lakes passenger steamers, the stately *Eastland*, pointed east toward Lake Michigan less than a mile away. At 275 feet in length and only 38 feet in width, she looked like a smaller and narrower version of a big ocean liner. Designed to travel fast across the Great Lakes and shallow connecting waterways, the *Eastland* had a normal draft of only 14 feet, something one might not expect given the visual proportions of her high steel sides, knifelike bow, and narrow beam.

This morning the *Eastland*, along with a handful of smaller ships, was chartered for the annual Western Electric Company cruise and picnic. In all, 7,000 employees of the Hawthorn telephone assembly plant in Cicero, members of their families, friends, or anyone else who could be cajoled into buying a ticket for 75 cents were boarding the *Eastland* or one of the other chartered ships for the 38-mile trip to Washington Park in Michigan City, Indiana. At the previous year's event the *Eastland* departed late due to the time it took to board the large number of passengers; accordingly, this year she was the first to open her gangway and would be the first to head out onto the lake. Despite the early morning hour and a light misty drizzle, the temperature was a comfortable 70 degrees. Passengers by the thousands now lined the wharf or were making their way by streetcar or foot over to the river.

The annual Western Electric picnic was a grand affair, and everyone dressed in their finest summer attire. The excursion was particularly popular with young adults and families, but especially among the many single young women who worked at the plant. It was the best opportunity all year for these hardworking and spirited Americans to socialize and meet people. Their names stumbled from the lips like roll call on Ellis Island: Marie Adamkiewicz; Florence Begtschke; Frantisek Danek; and Marie, Joseph, Rose, and Anna Dolejs, ages 17, 18, 19, and 20; Darowski, Dawska, Eicholz, and Erkman; Knoftz, Kupkowski, Kzarburg, and Landsiedel; Lane, McGlynn, McLaren, and McMahon. There were Hoffmans, Hipples, and Johnsons by the dozen. They were switchboard operators, inspectors, coil winders, laborers, drivers, foremen and forewomen, shop hands, shipping clerks, telephone assemblers, typists, and machinists, all decked out in their finest outfits, all ready for a fun-filled day off from work, all overflowing with life and optimism.

❖❖❖

The most eager to board, particularly those wanting to ride to Michigan City on the *Eastland* instead of one of the other ships, arrived at the wharf soon after daybreak. The early morning hour and the wet weather did not dampen the participants' spirits in the slightest. Captain Harry Pedersen, who took command of the *Eastland* the previous season, had been on board since before midnight, and Chief Engineer Joseph Erickson had already finished breakfast. By 6:20 he had worked his way down to his station in the engine room. Pedersen and Erickson, like so many of the crew and passengers, were European immigrants who had built productive lives for themselves here in Chicago and its suburbs.

By 6:30 the crowd on the banks of the river had grown to 5,000, and by 6:40 the *Eastland's* aft starboard gangway was open for the loading of passengers. The *Eastland* had five functional levels for equipment, crew, cargo, and passengers: (1) the lowest level, below the water line, with the engine room, boiler room, and crew space; (2) the Main Deck, just above the water line, containing the gangways, passenger areas, bar, galley and baggage and cargo areas; (3) the Cabin Deck, with more passenger space, a dining room, and the officers' mess; (4) the Promenade Deck, with additional passenger space and the smoking room at the stern; and, finally, (5) the exposed Hurricane Deck at the very top level of the ship. The whole affair was topped off with two large smoke stacks, one behind the other, in the middle of the ship.

The aft starboard gangway through which passengers were now boarding at a pace of about 50 per minute was slightly lower than the wharf, and passengers had to walk down steps from La Salle Street and then across the downward-sloping gangplank across the water and over to the gangway in the side of the ship. The gangway threshold was about four feet above the water line, but it could sink to one foot above the water when the ship was fully loaded. A waist-high dutch door on the aft gangway could be closed once the ship was under way, but was now open. Members of the crew collected tickets from passengers and kept a running count as they came aboard.

As passengers filed inside the enclosed Main Deck they tended to move upward to the higher decks, well above the water. As on most ships, there seemed to be more fresh air up top, and it was increasingly more open and easier to see out the higher one climbed. The Cabin Deck had far more portholes through her metal walls than did the lower Main Deck and was preferable in this regard. But unlike the Cabin Decks on other Great Lakes steamers where passengers could stand at an open rail, the *Eastland's* Cabin Deck, like the Main Deck, was fully enclosed within the exterior steel walls of the ship. A steel wall with portholes made it difficult to take in the view, so many passengers continued up to the Promenade Deck which had a walkway all the way around the perimeter of the ship. Plus, a five-piece orchestra was playing music for dancing on the Promenade Deck. The Hurricane Deck, one flight up, provided the best view of all, but it held the new lifeboats and other lifesaving gear and therefore had limited space for passengers.

Having an outside view or being on the upper decks was not good enough for many; some people also wanted to be on the right, or starboard, side of the ship, next to the wharf to watch those coming aboard or look out for arriving friends or family. A bit later on, at 7:30, when the ship cast off, passengers fortunate enough to be on the starboard side could make the customary goodbye wave to those staying behind on shore. It was part of the fun of traveling on a big ship. As a consequence of the movement of passengers both upward and predominantly to the right side, the *Eastland* developed a noticeable list — or unintentional tilt — to starboard, toward the wharf, within minutes of the first passengers boarding through the aft starboard gangway at 6:40. An additional 50 passengers with a combined weight of 3 to 4 tons came on board each minute.

❖❖❖

This list to starboard during boarding might seem very peculiar for such a large ship, but listing was not that unusual for the *Eastland*. She was designed with a very shallow draft, but also built to take on water ballast should the need arise. Down in the engine room below the water line, Engineer Joseph Erickson wrote in his log at 6:48 that a list to starboard had developed, and he ordered his engine room crew to right or steady the ship — to bring her straight upright again. "Boys, steady her up a little," were his exact words.

"Steadying her up a little" was done with the water

ballasting system — a series of very large water tanks, lines, valves, and a pump to move water in and out of the space in the lowest few feet of the ship from bow to stern. The system allowed the ship to be trimmed when under way to improve performance, but also enabled the crew to compensate for an uneven distribution of the load, especially a list to port or starboard. Like a moving bicycle on land, once under way and at speed on the lake, the *Eastland* usually straightened up nicely and moved comfortably across the water. She had developed a reputation, however, for being a bit "tipsy," especially while boarding and unboarding passengers. The water ballast system was the method for dealing with the issue.

An inlet below the water line on the port side ran to a longitudinally mounted manifold, about 12 feet long, near the floor in the center of the engine room. Pipes from the manifold fed the various ballast tanks, and the flow of water into each was controlled with valves, also located on the manifold. An air vent ran vertically from port and starboard tanks up to 10 feet above the water line. More than one tank could be filled at a time, but only a single detachable 16-inch handwheel was used to open and close all of the valves — an economical system, no doubt, but one not conducive to quick response by the engine room crew.

Although the tanks were filled using existing water pressure and gravity (because they were below the inlet and water line), emptying the tanks required the use of a pump in addition to the same manifold, valves, and pipes used during filling. With suction from the pump, water from the ballast tanks traveled back through the manifold and out an overboard discharge above the water line on the starboard side. This common inflow and outflow system meant that filling and discharging of tanks could not occur simultaneously. Also, it was not possible to shift ballast quickly by pumping water from one tank to another.

Instead, a filled tank had to be emptied under a discharge configuration, and then the other tank filled under a filling configuration — all using the single, detachable 16-inch handwheel to operate the assorted valves on the manifold.

Each tank had a vent through which a dipstick could be inserted to measure the water level, but a more common and faster method was used when loading from scratch at the dock: all of the tanks were emptied completely before any one or anything came on board. With the inlet valve open, each tank, depending on its capacity, filled within 10 to 25 minutes. The smallest tanks held 40 tons of water; the largest, 82.5 tons. The entire capacity of the tanks was 647 tons, and they could all be filled completely in about an hour. There were no flowmeters available, so Chief Engineer Erickson gauged tank levels by keeping track of how long valves were kept open. Each tank's vent pipe also gave off a characteristic whistle as it filled and the air in the tank was displaced under pressure; when it stopped or when water sprayed out the top, Erickson knew the tank was full and ordered the corresponding valve to be closed by one of the engine room crew using the interchangeable 16-inch handwheel. As with the filling procedure, the rate of discharge was also gauged primarily by keeping track of the amount of time the water flowed. The discharge pump was powerful, and it was possible to empty the tanks in equal or less time than was required to fill them.

Operation of the system required additional feedback, and Erickson had a number of sources of information in this regard. An 18-inch steel inclinometer mounted to the wall of the engine room was his main source of feedback about the ship's left-right list. However, rather than displaying the list in degrees, as one might expect, the tip of the pendulum-like inclinometer pointed to increments marked in inches; Erickson had to note each reading and calculate the actual degrees of list by hand.

Communication between the engine room and bridge was through a standard engine order telegraph containing the traditional small set of engine orders (e.g., stop, full speed, etc.) used on Great Lakes steamers. Any communication between the bridge and engine room regarding the operation of the ballast system occurred through a voice tube. Operation of the ballast system was traditionally under the full control of the Chief Engineer, however, and no communications concerning list took place between the bridge and engine room this Saturday morning. Erickson's management of the ballast system was based on the scheduled boarding and disembarking times, the time of day, planned loads, any information that might be shouted down to him from another crew member, and his own visual and kinesthetic sense of the state of the ship.

When Erickson ordered the filling of port ballast tanks at 6:48, the increasing load and list to the right had brought the aft starboard gangway through which passengers were boarding to within 18 inches of the Chicago River. It had been four feet above the water only a few short minutes before. Within a minute, however, the effect of the water flowing into the port tanks became noticeable and the *Eastland* slowly began to straighten up as weight was added to the left ballast tanks to compensate for the disproportionate number of passengers on the right side of the upper decks. The steel inclinometer on the wall of the engine room crept across the scale. At 6:51, only three minutes after action had been taken, it pointed to "0"; the *Eastland* was perfectly upright and on even keel once again.

❖❖❖

But unknown to Chief Engineer Erickson, being on an even keel did not mean the ship was stable or even remotely safe under the current circumstances. In actuality, the *Eastland* was in a most precarious and temporary unstable equilibrium. She was, to put it simply, in grave danger of rolling over like an egg in the water even though only about a quarter of the 2,500 passengers she was to carry were now on board. To understand the reasons why this was so and how this had all come about, one had to consider the history of the ship and, most importantly, recent "improvements" to enhance her safety in the wake of the sinking of the *Titanic*. Although Chief Engineer Erickson was certainly aware that modifications to the ship had been made during the preceding months, he most certainly was not aware of their full nature, their impact on the ship's stability, and their impact on his own life.

The *Eastland* was built in Port Huron a dozen years earlier, in 1903, by the Jenks Ship Building Company and designed to carry both freight and passengers between Chicago and South Haven during the spring, summer, and early months of fall. The primary freight was fruit, fresh and well-packaged for eventual sale in Chicago, secured in the lower levels of the ship. Passengers were to be carried in mid and upper decks. Although Jenks had a history of building reliable cargo ships for the Great Lakes, the *Eastland* was the first passenger ship the company ever laid down. She was also the last; Jenks got out of the ship building business in 1906 to become an automobile parts manufacturer.

During her first year on the Great Lakes, the *Eastland* showed no signs of instability or crankiness. She plowed her routes and delivered her cargo and passengers in a timely fashion, if only at slightly slower speeds than originally hoped. The first of her many modifications was carried out in 1904, at which time heavy equipment was rearranged to reduce her draft and speed her up to over 20 miles per hour. A heavy air-conditioning system was added for the Cabin Deck. She also became a nearly exclusively passenger ship, as opposed to the original passenger/freight configuration. The consequence of these changes, especially the movement of heavy machinery and the new air-conditioning equipment above the Cabin Deck, was that she was now somewhat top-heavy. In the parlance of naval architecture, shifting weight upward lowered her metacentric height (the measure of lateral righting moment, a boat's tendency to right herself once tilted left or right). A negative metacentric height is below the water line, a metacentric height of 0 is at the water line, and a positive metacentric height is above the water line. The higher the metacentric height, the higher the righting moment and the ship's inclination to right herself after an induced list. A large sailboat with a deep keel full of lead ballast has a very high metacentric height and a strong righting moment. A telephone pole floating in the water has a metacentric height of 0 or less and a weak or nonexistent righting moment. When first built, the *Eastland's* metacentric height when fully loaded was approximately 18 inches above the water line — typical for the shallow-draft cargo ships that traveled the Great Lakes but less than the metacentric height of two to four feet found on typical passenger ships where the human "cargo" could, unlike boxes of fruit, be expected to move about. The changes made in 1904 lowered her metacentric height from the original 18 inches.

In the 1904 season, after the first major modifications, a preview of later events occurred during a trip when thousands of passengers crowded the Promenade and Hurricane Decks. It was a very hot day, and passengers escaped the uncomfortable conditions in the enclosed decks by moving to the higher open decks. The *Eastland* developed a severe list and would most certainly have rolled over had the crew not instructed passengers to move downward to the lower decks. The owners subsequently limited the number of passengers allowed on the Hurricane Deck, and the *Eastland's* licensed capacity was lowered from 3,300 to 2,800. This event did not stop further modifications and the subsequent lowering of her metacentric height, however. In preparation for the 1905 season, a pair of large lifeboats were added to the Hurricane Deck, bringing the total to four such pairs of boats up top. Each lifeboat weighed nearly four tons. The listing problems continued, but the only response was to lower her licensed passenger capacity from 2,800 to 2,400 for the 1906 season. This was subsequently lowered to 2,200 and, later, to 2,000 for the 1913 summer season. Some cabins were removed prior to the 1914 season, giving her and additional 3,500 square feet of deck space. Her inspector, reasoning that she now had more room for passengers, increased her licensed capacity to 2,045; realistically, however, this would only make the *Eastland* more top-heavy if the ship were fully loaded with people.

On April 14, 1912, the *Titanic* struck an iceberg and sank in the northeast Atlantic on her maiden voyage. The unusually slow progress of her sinking resulted in the near-term survival of all passengers, an outcome contrary to general experience on sinking ships at the time and the rule of thumb that a third of all passengers most likely would never reach lifeboats before a large ship went down. Although the *Titanic's* lifesaving "boatage" and "raftage" capacity exceeded the requirements of the day, there was insufficient lifeboat capacity for everyone on board, and 829 passengers and 694 crew died as a result. The ensuing public outcry focused almost exclusively on lifeboat capacity and a call for "lifeboats for all." Naval experts correctly pointed out that passenger liner sinkings were actually quite rare, and when they did occur the angle of the ship and limited available time would render the loading and launching of such a large number of lifeboats a near impossibility, especially if thousands of people were involved. As a case in point, the *Empress of Ireland* sank off the coast of Canada in May of 1914 after being rammed by another ship, a catastrophe that rivaled the *Titanic* disaster. *Empress of Ireland* went down in just 14 minutes, and having twice as many lifeboats on board would not have reduced the loss of life. The cause of safety would be better served, many experts stated in response to the lifeboats-for-all movement, with improved communications, enhanced ship operating performance, and tracking ships and the occasional rogue iceberg.

But "lifeboats for all" it was to be. On March 4, 1915, President Wilson signed into law the *La Follette Seamen's Act* requiring boatage and raftage space for all passengers and crew aboard American ships. Owners of ships operating on the Great Lakes were given one year to comply. The *Eastland*, as it turned out, had been sold the previous year to new owners, The St. Joseph-Chicago Steamship Company, who failed to delve fully into her "cranky" history. In preparation for the 1915 summer season and with an eye on the compliance date for the new federal regulations, they scheduled further "enhancements" to the ship. First, two inches of concrete were laid down in the dining room on the Cabin Deck to repair a rotting wood floor. At 150 pounds per cubic feet, this added 14-19 tons to the ship — about a dozen feet above the water line. The floor on the Main Deck where passengers entered was also rotting, possibly from water spilling in through the low gangways, and an expansive layer of concrete was applied here as well, bringing the total for both areas to 30-57 tons.

In addition to requiring "lifeboats for all," the impending enforcement of the *La Follette Seamen's Act* would regulate the number of passengers ships could carry based on current capacity and other factors. The St. Joseph-Chicago Steamship Company and Captain Pedersen were aware of this and had calculated that the *Eastland's* licensed capacity was going to be reduced for the 1916 season to somewhere between 1,028 and

1,552, depending on the future inspector's interpretation of the new regulations. This would be a substantial reduction in her current licensed capacity and would make the *Eastland* unprofitable to operate. If they could somehow increase the capacity to 2,500 for the 1915 season, they reasoned, the less they would actually lose when the law came into effect. Accordingly, Captain Pedersen had three large lifeboats and two life rafts moved to the *Eastland* from another ship and acquired four additional life rafts. The *Eastland* was launched in 1903 with 6 lifeboats. She now had 11 lifeboats, 37 life rafts, and a work boat, all of which were mounted on the Hurricane Deck, and 2,570 heavy lifejackets stowed on the Hurricane Deck, Promenade Deck, and crew's quarters. In addition, heavy railings made of gas pipe were installed on the Hurricane Deck to keep passengers away from the visually prominent lifesaving gear. At no time was the ship's center of gravity or metacentric height ever calculated or tested. The inspector approved the ship for a capacity of 2,500, as the owners requested, and the *Eastland* reentered service on July 2. For the next three weeks, until July 24th and the Western Electric excursion, she operated with small passenger loads.

❖❖❖

Back in the engine room, at 6:53, just two minutes after having brought the ship to an even keel at 6:51, Chief Engineer Erickson saw the pointer on his inclinometer moving slowly once again, although this time it told him the *Eastland* was starting to tilt to port, away from the wharf. When it reached about 10 degrees, he gave his crew orders to open the valves on the manifold to fill the starboard tank. The engine room crew cranked the valves to the starboard tank open with the interchangeable handwheel and they stayed open for the next four to five minutes. The *Eastland* rolled slowly upright and the inclinometer once again hung perfectly vertical.

Certainly at this point, but probably minutes before, the *Eastland* was doomed. The modifications completed earlier in the month, especially the addition of the tons of new lifesaving gear added to the Hurricane Deck, had had a dramatic effect on the center of gravity and metacentric height of the ship. Erickson was never given any information which would or should lead him to believe that the metacentric height of the ship had been altered or that the maximum legal passenger capacity had been increased courtesy of the owners, the captain, and an obliging inspector. Although he had been told that they would have a full passenger load, it was Erickson's understanding that this would entail bringing 2,045 passengers aboard, not 2,500, a difference of perhaps 30 or more tons of unpredictable human cargo. Although Erickson believed that he was filling port or starboard ballast tanks to keep a stable ship on an even keel based on the movement of a few erratic passengers, in reality, the *Eastland's* metacentric height was now less than zero. She was as unstable laterally as a floating telephone pole. Erickson had managed to keep the ship from rolling over only through his methodical management of the ballast system.

Minutes later at 7:00 the passenger count at the gangway reached 1,600. The ship still favored a list to port, leading Erickson to conclude that passengers were congregating on the port side when, in fact, a preponderance of passengers were along the starboard side. He ordered the engines started at 7:05 to warm them up and at 7:07 ordered the engine room crew to start the ballast pump to remove water from the No. 3 port ballast tank. This action would replace the water ballast on the left side of the ship with air and should correct the continued list to port. It also meant, however, that starboard tanks could not be filled at the same time.

The informal count of boarded passengers reached 2,500 at 7:10, and those remaining in line were redirected to one of the other ships hired for the excursion. The list to port still persisted and, at about 10 degrees, had not improved despite the continued pumping from the No. 2 port ballast tank. Passengers generally found the listing — first one way and then the other — amusing and paid no attention when the radio officer went out on deck and asked people to move to the starboard side to help balance out the ship. The five-piece orchestra struck up a tune at 7:13, and people started dancing on the Promenade Deck aft, further shifting the load upward. And in stark contrast to the 1904 event in which passengers on the Hurricane Deck were told to move to lower decks to prevent her from turning turtle, today the Hurricane Deck was outfitted with tons of lifeboats and related gear which, unlike cooperative passengers, could not be moved to lower decks on short notice.

❖❖❖

At 7:16, as David Durand and Walter Perry watched from their perch three stories up across the river and as the *Eastland* sat tied up to the wharf but leaning 15 degrees outward toward them, down in the engine room Erickson was giving orders for the ballast system to be reconfigured to bring water into the starboard tanks. Rather than continuing to pump water out of the port tanks, they would pump water into the starboard tanks. Erickson personally opened the valve for the No. 3 starboard tank, his brother Peter opened the valve for the No. 2 starboard tank, and another crewman opened the valve from the intake. A minute passed by during which time nothing seemed to happen or change. The *Eastland* sat frozen in her awkward, schizophrenic stupor. Suddenly, Ray Davis, an assistant to the ship's owners, burst into the engine room and asked Erickson if something was being done to straighten the ship. Erickson replied that they were doing everything possible, and he asked Davis to return topside to see if the starboard fender strake was hung up on the wharf, preventing the starboard side from rotating back down into the water.

One minute later at 7:18 the *Eastland* started to straighten up. It began slowly but then quickened, and at 7:20 the ship was as straight up and down as she should be. "I believe we are getting her," said Erickson to his engine room crew as the *Eastland* moved back toward the vertical, believing their attempts had been successful. In reality, however, it was miraculous that the ship was upright at all. But with the listing problem apparently solved and with no knowledge whatsoever of the ship's true instability, Erickson and the rest of the crew began to prepare for her departure from the wharf. The gangplank to the aft starboard gangway was brought in and the aft line to the wharf was cast off. In keeping with the tradition of waving "goodbye," a majority of passengers on the exposed upper decks lined the starboard rail rather than the port rail.

The *Eastland* stayed on an even keel for less than a minute. By 7:21 she started to lean to port once again. First 5 degrees, then 10 degrees, then 15 degrees, and 20 degrees. Erickson heard water, lots of water. It was spilling in through the scuppers on the port side of the Main Deck. He wondered if starting the engines had somehow upset the balance of the ship and ordered them to be shut down. But it had nothing to do with that. The list continued beyond 20 degrees. Erickson instructed one of his crew to run up topside to pass the word that passengers should move to the starboard side to counter the list to port. The message was spread through the crowd, and more passengers moved over to the right side of the ship.

The list to port continued, and at 7:23 the water began to pour into the port gangways and down into the engine room.

Captain Pedersen, up on the bridge, who later referred to the list to port at this moment as only a "trifle," signaled to Erickson on the engine order telegraph to "standby." The stern line was cast off, and the back end of the ship began to swing out slightly into the Chicago River in preparation for departure. As the ship drifted away from the wharf, passengers began to step away from the starboard rail, toward the centerline of the ship and toward the port rail, which was now down at an angle of nearly 25 degrees.

Incredibly, at 7:25 as her stern drifted out further into the river, the *Eastland* seemed to be righting herself again. For one minute she wavered in the balance, still tilting to port, but holding steady. But all hell had broken loose in the engine room. Water washed across the floor. Erickson issued orders for his crew to start the bilge pump. A minute later she stopped her rotation towards vertical and started tilting to port one last time. Experienced crewmen knew what was about to happen. The oilers and stockers bolted for the ladders as the list to port passed 30 degrees.

Up on the Promenade Deck the passengers were asked to move to the starboard side again, but the wooden floors were now too steep and slippery from the drizzle for anyone to climb upward. Dancing was no longer possible, so the musicians dug in their heels and switched to ragtime. Passengers laughed and talked, incredibly still not cognizant of the danger. The *Eastland* continued listing further to port, and at 7:28 her decks slanted an unbelievable 45 degrees. A piano slid off its supports on the Promenade Deck and a large refrigerator in the bar crashed and slid across the floor. There were a dozens screams and terrified shouts. Water poured in through the port gangways and Main Deck port holes. Passengers on the Main Deck rushed to the stairway. Down below in the engine room Chief Engineer Erickson turned on the water injectors to cool the boilers, fearing they might explode when hit by the river water.

From their third-story perch above the river, David Durand and Walter Perry watched as the *Eastland* slowly rolled over onto its side without a splash. In an instant thousands of people were in the river, screaming and thrashing, grabbing in panic for anything, including each other. The *Eastland*, lying flat on her side, settled to the muddy river bottom, her port half underwater and her starboard half above. The two warehouse employees bolted down the stairs and over to the bank of the river to help.

❖❖❖

When all the bodies were finally retrieved, identified, and counted, the final death toll was 844 passengers — 15 more passengers than died on the *Titanic.* With few exceptions, those who survived were on one of the upper exposed decks when she rolled over, and those who died were inside the enclosed Main and Cabin Decks. The lifeboats remained attached to their davits, the lifejackets packed in their boxes. Like all 70 members of the crew, Chief Engineer Erickson knew the ship well enough to know what to do. As the water reached his neck, he slipped out of the engine room along a steering cable under the main deck, slithered through an air duct and over to an air pocket under a port hole on the starboard side where he was pulled out to safety.

❖❖❖

After twenty years of litigation, on August 7, 1935, The United States Circuit Court of Appeals upheld a lower court ruling that the owners of the *Eastland*, the St. Joseph-Chicago Steamship Company, were not responsible for the disaster. The ship was ruled to be seaworthy and "all proper precautions" had been taken by the operators. Blame was conveniently placed on Chief Engineer Joseph Erickson who, according to the court, was negligent in his duties to fill the ballast tanks properly. Erickson was not present to defend himself — he died of heart failure in 1919 at the age of 37.

REFERENCES AND NOTES

Bonsall, T. E. (1988). *Great shipwrecks of the 20th century.* New York: Gallery Books.

Chicago historical information, 1915, July 24: Eastland disaster photos (undated). *Chicago Historical Information www site.*

The Eastland disaster [video] (1999). Kenosha, Wisconsin: Southport Video.

Hilton, G. W. (1995). *Eastland: legacy of the Titanic.* Stanford, California: Stanford University Press.

Ratigan, W. (1977). *Great Lakes shipwrecks & survivals.* New York: Galahad Books.

Timeline of the morning of the disaster (2001). *Eastland Disaster Historical Society www site.*

Watson, M. H. (1987). *Disasters at sea.* Wellingborough, U.K.: Patrick Stephens.

DRIVEN TO DISTRACTION

The highway outside Marseilles twisted around the rugged mountains inland and high above the Mediterranean coast. It was a driver's road, the kind that played back to you when you steered confidently up through smooth curves and accelerated out and then down inviting long downhill stretches until you braked firmly again and set up for the next sweeping turn and the reward of another unwound ribbon of asphalt even more perfect than the last.

April was the best time of year in Provence. The cold winds of winter were gone, the hordes of foreign tourists had not yet arrived, and August and the invading army of vacationing Parisians was many months in the future. It was a beautiful day. All one had to do while driving this particular day was mind the road and the many groups of bicyclists in their colorful outfits out plying the rural highways at the opening of the cycling season.

❖❖❖

The driver's name may have been Monique, which is as good as any other name considering that the police spokesman did not give her name when he announced to the press that she had been arrested later after the carnage on the highway earlier that Sunday. He did say that she was 27 years of age. Her companion in the passenger seat up front was clearly involved but remained unnamed as well and was not the person responsible for operating the car. Yes, it was appropriate in this case that the finger be pointed directly at the one whose foot was on *l'accélérateur* and whose hands were not entirely on the steering wheel and whose eyes were someplace other than the road ahead. There was also, the policeman said, a contributing factor, an unusual mitigating circumstance, if you will, involving Monique and her companion that played a role in the accident. Monique's mind and hands were elsewhere for many critical moments as she sped down the scenic highway across the rolling French countryside.

It was not a matter of the speed or the road; it was all about *the matter of the distraction* which came when she least expected it. One might have guessed it was a bit of fooling around or even an incoming call on a mobile phone, but neither was the case. It was something else that was inside the car.

❖❖❖

Everything was pleasant and controlled one second before, but the next second the shrill *beep beep beep beep beep beep* distress signal blared out from somewhere around the dashboard and filled the cabin. Panic set in.

"What is it? What is it?" shouted the companion in the passenger seat, wondering what the sound was and from where it was coming. The pure-tone distress signal was very difficult to localize.

"The Tamagotchi!" came the frantic reply.

"The Tamagotchi?"

"Yes! The Tamagotchi!"

The key in the ignition was attached to a small chain which also looped through the small egg-shaped device dangling next to the steering column. It was Monique's Tamagotchi, an imported object from Japan. As the manufacturer said, the Tamagotchi was "the original virtual reality pet," a flattened colorful egg with three tiny buttons across the bottom, a small liquid crystal display in the center, and a key chain attached to the top. At the level of reality — not virtual reality — the Tamagotchi was a miniature video game which received inputs from the player via the three little buttons and provided feedback with moving pictures on the display and piercing auditory beeps when it needed serious attention from its owner.

Monique had acquired her Tamagotchi some days before and knew that the objective of the whole thing was to keep the Tamagotchi "alive" for as long as possible — just like a pet dog or cat. Each Tamagotchi was, according to Bandai Company, the manufacturer, an egg from distant space that had landed on Earth. Once actuated or "woken up," it required "nurturing" and "parenting" from its human parent. Its "lifespan" depended on how it was cared for, and it signaled its need to be "fed, nurtured, and cleaned" through its display and beeps, to which the caring parent had to respond in a timely manner by pressing the three buttons to provide "food" or "attention." It took a new owner quite a while to figure all these details out, and sometimes it took a seemingly random approach to pressing the tiny buttons with the very tip of your finger to get the beeping to stop and the Tamagotchi back to a healthy state as shown on the tiny display.

Once "hatched," it could not be turned off and therefore required near-constant attention, especially during its early days of life. "It is not a game," said Tomio Motofu, the spokesperson for Bandai. "You're looking after a space creature whose lifespan depends on how you care for it." "It is more than a toy," added Mary Woodworth of Bandai USA Division, "it is a learning device. It teaches people to be responsible."

❖❖❖

And at this moment Monique took her responsibilities — at least the responsibility of keeping her Tamagotchi alive — very seriously, more seriously than her responsibilities as the driver of the car. The Tamagotchi's bleating beeps meant that it was in very critical distress and that it required immediate and undivided attention in order to survive. At most, a Tamagotchi could live for 26 days, the current record in Japan. Anything over 17 days was rare. But Monique had not had her Tamagotchi nearly that long — only a matter of days, actually — and she would be considered a failure as a parent should her Tamagotchi expire prematurely as a result of her lack of attention and poor parenting skills.

The Tamagotchi on the key chain dangling from the ignition of the car rolling down the highway with the cyclists in the hills of Provence was indeed expiring. Its beeps were frantic and loud, the kind of sound heard only when a Tamagotchi was near death due to neglect early in its short life. Its buttons had to be pressed and its little display examined closely to see what was wrong and what it needed to continue to live.

"The buttons! The buttons! Press the buttons!"

Her passenger now understood that this all had something to do with the round electronic gadget on the key chain and leaned forward and grasped it, being careful not to pull the attached key out of the ignition of the running car.

"You must hurry! Press the buttons! The Tamagotchi is in distress!"

But this was all so unfamiliar to the companion who was fiddling with the Tamagotchi, looking for the tiny buttons so they could be pressed to somehow save the life of the egg from space. It was taking too long, and Monique took her eyes from the road and a hand from the wheel and pointed to the

Tamagotchi and said that it was "here" that the buttons had to be pressed while the car drifted away from the center of the lane, then precariously to the side and then near the shoulder. Her attention was now fully heads down, focused entirely on the dying Tamagotchi on the key chain hanging from the ignition, telling her friend in a panic what to do to save its life when the car, traveling three or four or five times faster than the pack of cyclists ahead on the side of the road, mowed into the back of the group from behind. One rider took the full force of the front of the car and flew up and out and over and down to the pavement with instantly fatal injuries; another was injured. The Tamagotchi expired as well.

REFERENCES AND NOTES

Driver saves virtual pet (1998). *Reuters News Service*, April 8.

NEGATIVE TRANSFER

NASA 1 [Flight Control, Edwards, California]: "Okay, cockpit camera on now."

Milt Thompson [Test Pilot, in the M2-F2 Lifting Body under the B-52 mother ship's right wing]: "Okay, coming on."

NASA 1: "All systems go, thirty seconds now, Milt."

Thompson: "Okay."

B-52: "On speed."

NASA 1: "Fifteen seconds now."

Thompson: "Roger."

NASA 1: "Ten seconds now."

Thompson: "Okay. Five, four, three, two, one, release!"

The silver wingless aircraft fell free from the B-52. Milt Thompson looked up for just a second through the Plexiglas bubble canopy over his head as the B-52's massive right wing receded into the sky above. The release was clean and surprisingly smooth, not at all like the hard launch when he was in the X-15. This was different. He had simply flown away from the mother ship without any lurching or jerks. "A pleasant surprise," he thought, and a relief considering that the M2-F2 lifting body had never before flown on its own in the air. First flights were always a little nerve-racking, but so far this had been a piece of cake.

Forty-five thousand feet of empty space now lie between Thompson and the Mohave desert below. He was currently pointed north. The sun had just come up over the horizon off to his right. The flight plan called for him to make a 90-degree turn to the left, followed by another left turn, test the peroxide rocket along the way, deploy the landing gear at the last possible second, and put the aircraft down horizontally on Rogers Dry Lake. If all went well, and even if it didn't, he would be on the ground in under four minutes. Either way, it was going to be a wild ride.

❖❖❖

The M2-F2 was the most peculiar-looking aircraft ever flown by NASA — if you could call a near free-fall from over eight miles up "flying." Officially, it was "a modified half-cone, rounded on the bottom and flat on the top, with a blunt rounded nose and twin vertical tail fins." Unofficially, some likened its shape to that of an old-fashioned footed bathtub with a tapered boatlike underside. The idea of a wingless lifting body had originated with Dr. Alfred J. Eggers Jr., a NASA R & D director up at the Ames Aeronautical Laboratory south of San Francisco. The central issue triggering his thinking was how to bring a manned spacecraft back through the Earth's atmosphere without parachutes or complicated splashdowns in the middle of the ocean. A controlled flight back from space was far more desirable than a "plunge to Earth in a ballistic trajectory." A flyable returning spacecraft could be landed on a conventional airport runway, and the craft might even be reused for multiple return trips into space. But there were many problems associated with a "flyback" return, the greatest being that conventional aircraft wings and any control surfaces attached to them would melt away. Eggers reasoned that a lifting body shape with a blunt nose, round bottom, and flat top would provide sufficient lift to enable a pilot to actually steer the craft down from Earth orbit to a runway and survive the extreme heat of reentry.

❖❖❖

Milt Thompson had been at his desk at the pilots' office a few years before in 1962 when he heard NASA engineer Dale Reed running up and down the hallway outside. Reed had been advocating the construction of a flight test model of a lifting body, and now he had taken matters into his own hands in a very clever way. He had carved a small lightweight model of a lifting body and attached it to a long string. The longest enclosed straight space around just happened to be the hallway outside the pilots' office, and Reed tested his design by running up and down the floor, his little lifting body model in tow. To the surprise of just about everyone present, the model was very stable in its simulated flights down the hall. Reed's antics were more than a test, however; he was also floating his proposal that the center undertake a relatively small-scale program to develop and fly a piloted lifting body. Initially, at least, they could test a prototype by towing it rather than investing in an expensive propulsion system. Reed was up on the roof launching his model out towards the aircraft ramp after his demonstration in the hallway to further his case and appease his growing cadre of followers, including a number of test pilots.

Reed's greatest obstacle was available resources at the center, not lack of interest. The X-15 research program was in full swing and the director, Paul Bikle, was not about to divert engineers, mechanics, and money to Reed's proposed program. But Reed was persistent and he moved his desk, along with his drawings, charts, models, and recently recorded movies of his lifting body models in flight, into the pilots' office. In no time, Reed had worked his charms on Milt Thompson, hooking an advocate and skilled test pilot at the same time. Reed,

Thompson, and an engineer named Dick Eldredge brought Eggers down from the Bay Area to obtain his endorsement and presented a plan for a six-month feasibility study. Bikle could no longer swim against the changing tide of interest in manned lifting bodies, and he approved the program. Progress was rapid, and in September the small team and a sailplane contractor started work on the M2-F1, a full-size flight-capable lifting body with a wooden exterior. Bikle, having now fully endorsed the concept, became the program manager. Thompson had become the principal test pilot. By March, 1963, the M2-F1 was being towed on a 1,000-foot line behind a '63 Pontiac Catalina convertible with a 421-cubic-inch triple-carb Tripower engine, more than enough horsepower to get the M2-F1 and Thompson up into the air at 120 mph. By early 1964, they had installed an ejection seat and were flying the M2-F1 at altitude, pulled aloft by a C-47. There had been many accomplishments and many successful test flights during those few short years, and many opportunities to celebrate late in the afternoon at the Rock-a-Bye saloon in Rosamond out beyond the edge of the dry lake.

❖❖❖

The initial test flight of the M2-F2, now under way during the still of daybreak, July 12, 1966, was an important one for the lifting body program. As the first of the "heavyweight" designs, the M2-F2 was a serious aircraft made of aluminum and steel, not wood and resin. It weighed in at a hefty 4,620 pounds, without ballast and test pilot, and was 22 feet long and 10 feet wide. Milt Thompson sat forward in the craft, his head literally poking up above the flat top but under the bulbous Plexiglas canopy. Down beyond his feet was the rounded Plexiglas nose through which he could see the ground below.

The F-104 chase plane slightly behind and to the side chimed in on the radio. Milt Thompson's flight was progressing well: "It's a beauty."

NASA 1: "Heading is good."

Thompson: "Okay."

NASA 1: "Coming up on forty thousand feet now; check alpha [angle of attack] and airspeed. And, start your turn."

It was now a scant 26 seconds since Thompson had released from the B-52. He had fallen one mile closer to the desert floor. But with a forward speed of 450 knots at release, he had covered more than twice that much distance over the ground. At top speed and under the best of circumstances he could traverse three miles for every mile he fell. His first major maneuver was the first 90-degree turn to the left to set him up headed west, running along the north shore of Rogers Dry Lake. He pushed the stick to the left and the M2-F2 started a smooth westward turn.

The faster he went the more lift the M2-F2 would generate, and he needed to pick up speed and lift if his landing on the lake bed was to be horizontal rather than vertical. Thompson pushed the stick forward to tilt the nose down and start picking up airspeed. The next maneuver in the flight plan was a practice landing flare at altitude in which he would fly the M2-F2 relatively fast and then pull back on the stick to level her out as if he were landing.

Thompson pushed over and immediately began to pick up speed. Simultaneously, he felt a slight left-right oscillation, as if his bathtub-shaped aircraft was rolling a bit from side to side, like a baby's cradle being rocked one way and then the next. Aerodynamic simulations had predicted that this instability might occur during flight, and he knew he had to be exceptionally careful about introducing a pilot-induced oscillation (PIO) in the M2-F2, particularly during a turn. The shape of the lifting body could make recovery from a PIO very difficult — or even impossible — to handle.

❖❖❖

NASA and Northrop (the aerospace contractor selected to build the M2-F2) faced one particularly unique challenge when designing the control system for the M2-F2 some months before. Unlike the M2-F1, which had short horizontal tip fins and elevons, the M2-F2 had no horizontal wing surfaces which might be used also for mounting moveable surfaces to control roll. Wings would project out into the free stream flow during reentry, and everyone wanted the M2-F2 lifting body design to be as close as possible to a real spacecraft that would have to withstand the extreme heat when entering the atmosphere at 17,000 miles per hour. How, then, were they to make the craft roll left and right, which would be necessary when turning or leveling out to land?

The decision was to eliminate the elevons entirely and use the other moveable surfaces to control roll. There were only two: the left-right inboard ailerons at the rear and the left-right vertical tail rudders. Wind tunnel tests showed that using the left-right inboard ailerons alone to control roll would not work. Lifting one aileron produced a high-pressure region above it and inboard of the adjacent vertical fin. The result was a dangerous yawing moment and, surprisingly, a roll moment in the exact opposite direction of the original control input! No one ever said flying a bathtub would be easy.

The solution was to interconnect the control of the rudder on the two tail fins and the two parts of the horizontal aileron. When the pilot made inputs to move the ailerons, the rudders would move proportionally to counteract the yawing moment produced by the moving ailerons. Furthermore, Thompson could change the degree to which the rudder moved in concert with the ailerons by adjusting a control, a vertical hand wheel off to his left that they called the *interconnect ratio changer.* Roll it one way and you got a large response from the rudder with an aileron input; roll it the other direction and you got next to no response from the rudder in concert with your aileron input. The adjustment was necessary for two reasons: first, no one was quite sure what ratio would work best and, second, Milt was likely going to need to change the ratio as he changed speed and moved through different layers of the atmosphere with their different densities. They all knew that it was not the ideal way to solve the problem, but it seemed like it would work.

On the downside was the fact that the aircraft was highly sensitive to pilot-induced oscillations, in which moving the stick to the left would result in the aircraft rolling to the left — but perhaps further than intended by the pilot. The pilot would invariably respond by moving the stick to the right, and the aircraft would roll hard to the right and so on, back and forth, with each cycle becoming more and more pronounced until he lost complete control. If the setting on the interconnect ratio changer was just right, however, he should be able to control everything and avoid the pilot-induced oscillations.

Milt had spent hour upon hour in the M2-F2 flight simulator going over the subtle handling characteristics of the craft and the flight plan. Using the Air Force's X-15 simulator cockpit, they had been able to simulate not only the M2-F2 controls, but also the handling characteristics of the M2-F2 under all conceivable circumstances. Just as he had shown by his skilled handling of the M2-F1 while being towed behind the Pontiac, Milt's skill flying the M2-F2 simulator impressed everyone on the team.

❖❖❖

With the M2-F2's nose now angled down, the airspeed

indicator passed 290 knots. Thompson's altitude was now down to 22,000 feet, over halfway there. He pulled back gently on the stick, bringing the nose up level, and then up a little higher, simulating the landing he planned to make once he reached the level of the lake bed. The flare maneuver lowered his rate of descent and also burned off some speed. The g forces built up to 1.5 times the force of gravity as he slowed down, the nose still angled up somewhat. At 18,000 feet his speed had fallen back down to 200 knots. He brought the nose down level again. So far, so good.

The next step in the tightly orchestrated flight plan was to fire the XLR-11 peroxide rocket in the tail, the same type of rocket used in the Bell X-1 flown by Chuck Yeager. The purpose of the rocket was to extend the M2-F2's range by providing thrust. With the propulsion provided by a rocket, he could cover a greater distance and descend at a slower rate. This could come in handy during landing or if there were an unexpected change in plans during the "glide" down to Earth. But if the rocket was not aligned perfectly with the center of gravity of the aircraft, he could careen off in some insane direction and break up. Thompson made doubly certain that the nose was straight and level. He raised the safety cover and flipped the switch. The rocket kicked in and he accelerated straight ahead over the desert as planned.

❖❖❖

Seconds later he was setting up for the final turn and landing on the forgiving expanse of dry lake bed. Ground control came on to help him stay exactly on the plan.

NASA 1: "Okay, add one degree on your upper flap."

Thompson: "Okay."

NASA 1: "Okay, do you have the field in sight?"

Thompson looked out forward and to the left through the canopy, across the Mohave Desert. He was headed west and the rising sun was now behind him. The marked landing area on the lake bed was many miles away, but in clear view to the south. "Affirm."

"Check your dampers and interconnect and start your turn anytime. Start your turn, Milt," said NASA 1.

"Okay," replied Milt Thompson. He cautiously moved the control stick to the left to initiate the turn.

❖❖❖

It was at this juncture that his problems began. The M2-F2 lifting body did not respond quite as planned. Flight controllers on the ground were receiving near-instantaneous flight data from M2-F2. Everyone on the team was aware and focused on the unfolding drama. For Milt Thompson, it was the rapidly changing view out his little cockpit; for those on the ground it was the shocking, sudden seesaw movement of an automated ink pen on the chart recorder. Yet no one said a word. It would only distract him.

As Milt Thompson passed through 16,000 feet on his final turn to the marked landing spot, he sensed that the sensitivity of the controls was too great. The tiniest of inputs to his stick made the M2-F2 oscillate slightly from side to side. His gaze alternating between his instruments and the view outside, he reached down and felt for the interconnect ratio changer and moved it slightly to decrease control sensitivity. That seemed to help, but, still, he sensed that he was on the edge of a controllability problem. He reached back down with his left hand again and moved the interconnect ratio changer once more, simultaneously coordinating the movement of the stick in his right hand and the rudder pedals with his feet. His turn nearly complete, he pushed the nose over to pick up speed, headed for the dry lake bed ahead.

Suddenly, just as he had feared, the vehicle began to roll, first to the left and then to the right, just 5 degrees or so to each side. Each time he moved the stick to counteract the roll in the opposite direction he found himself rolling back even harder in the other direction. He reached for the interconnect ratio control again, moving it to further lower the sensitivity of the flight controls, but the controls seemed to become even more sensitive. Now he was rolling 10 degrees to one side and suddenly 10 degrees to the other.

M2-F2 was now in a 30-degree dive, driving toward Rogers Lake Bed 8,000 feet below at a rate of 300 knots. Thompson's rate of descent was 18,000 feet per minute. If he did not get M2-F2 under control he would pound into the hard clay bottom in 27 seconds. He grabbed for the interconnect ratio control and moved it all the way to the lowest setting. The control wheel slammed up against the hard physical stop. Suddenly, he was rolling a full 45 degrees to the left and quickly 45 degrees to the right, a full 90 degrees each second, one side to the other. The motions were violent, and his robin's-egg blue helmet slammed up against the left side and then the right side of the Plexiglas canopy each time the silver M2-F2 rolled wildly from one side to the other.

Milt Thompson was experienced enough to know that this was now a matter of life and death. He had do so something, but the question was "what?" Pilot-induced oscillation was just that — pilot-induced. If he made no control inputs there should not be any pilot-induced oscillations. If he centered the stick, that would still be a control input, wouldn't it? The lifting body was supposed to be aerodynamically stable all on its own, just like Dale Reed's scale model that he pulled down the hallway and glided off the roof. It should work, he told himself. It should work. It had to work. "Let go of the stick, stupid! Let go!"

❖❖❖

The over-the-shoulder movie camera made a fitting record of the moment and this NASA test pilot with the right stuff. As the horizon rolled from straight up and down to straight down and up each second, as Milt Thompson's helmet banged against one side and then the other of the clear Plexiglas canopy, and as the desert floor came into focus as in a big magnifying glass, Milt Thompson's black-gloved hands sprang back and away from the control stick, just as if they had been burned. It was his only way out and his last hope.[4]

❖❖❖

The M2-F2 lifting body, a 4,620-pound silver bathtub plummeting through the sky at more than 300 miles an hour, magically stopped oscillating from one side to the other and leveled out, stable as ever. Milt looked down at the interconnect ratio control and saw that it was set to the *maximum* sensitivity position. He had been moving it in the wrong direction! He also realized in that moment that the interconnect ratio control in the simulator was a lever, not a wheel, and that it moved in exactly the opposite direction as the actual control now under his left hand. He was the victim of a classic case of *negative transfer*. He reset the interconnect ratio control to its correct position, grabbed the control stick, pulled up to just above the level of the lake bed, slowed to 240 knots, dumped the landing gear, and sat the M2-F2 lifting body wheels down on the lake bed. A

[4] A Northrop flight control engineer watching the film during the M2-F2 postflight review nearly passed out.

magnificent roostertail of dust followed his track across the hard packed clay. Without a parachute or the wheel-less skids of the X-15 to slow it down, the M2-F2 would roll on southward across the desert for over a mile and a half. Milt Thompson enjoyed the ride.

NASA 1: "Beautiful, Milt."

B-52: "At least it goes a little farther than the X-15, doesn't it?"

Thompson: "Yes, it do. I'm trying to get to the Rock-a-Bye."

REFERENCES AND NOTES

NASA facts. The lifting bodies (undated). *NASA Dryden Flight Research Center www site.*

On the frontier: flight research at Dryden 1946-1981. Chapter 8-4: the "heavyweights" fly (undated). *NASA www site.*

Reed, R. D. (1999). Wingless flight: the lifting body story [online book]. *NASA Dryden Flight Research Center www site.*

Thompson, M. O. (1992). *At the edge of space: the X-15 flight program.* Washington, D.C.: Smithsonian Institute Press.

Thompson, M. O. and Pebbles, C. (1999). *Flying without wings: NASA lifting bodies and the birth of the space shuttle.* Washington, D.C.: Smithsonian Institute Press.

All dialogue in this story is from *Flying without Wings* (Thompson and Pebbles, 1999) and the M2-F2 test flight transcript.

END GAME

In the morning when the sun burned bright over the blue of the Aegean Sea and the thousand islands of Greece, he did not want to face another day of life. For Pandelis Sfinias, architect of a business empire, Chief Executive of the ferry company Minoan Flying Dolphins, Vice President of the parent company Minoan Lines, President of the Union of Coastal Ship Owners, multimillionaire Greek business executive well-connected to the most powerful politicians, it was not a day to savor the Mediterranean sun or the view out the sixth-floor office window over the industrial port of Piraeus on the outskirts of Athens. It was only another day to bear the weight, the angst, the thick depression of the two months since September 26, 2000 when the *Express Samina* ferry went down in the Aegean Sea. "These things happen," he told himself early on. They had a large fleet of ships and the law of averages would eventually work their way around if you stayed in the game long enough. "Everything would heal with time," consoled his friends. He had at least ten or even twenty years left in his career and they said he had to accept the past and move on.

But the game was not worth playing anymore, he said to himself under his breath, submerged below the depressants and alcohol still in his blood, sitting silently at the table during the morning meeting in his office suite high up in the building. He wanted no part of it, no part of anything anymore. What was the point? There was only one path out — out and away from the ceaseless avalanche of words and anger and threats and hatred that had washed over the whole of the country and the whole of his floating empire, Minoan Flying Dolphins. There was no option, but for the moment there was the matter of the meeting to attend to and the summary required by his guests in his office high above the street.

❖❖❖

It had not started out this way, he began to tell the visitors, sitting again at the table and projecting back in time. In the beginning only a decade before it was an ingenious scheme to unify and revive the large Greek ferry business, a sound business plan with important links to government and regulators to serve 10 million Greek ferry passengers each year, an arrangement to make a great amount of money for the key investors. What they had to do first was create a new company, Minoan Flying Dolphins, owned by the parent company, Minoan Lines. Next, they would buy up as many as possible of the smaller family-owned Greek ferry companies servicing select routes into and out of Greece from Italy and Turkey, but mostly the lines between the major ports sprinkled around the Aegean. Ferry boats, not airliners or bridges, were the veins and arteries of Greece, carrying passengers and cargo throughout the Archipelago as they had done for eons. Whoever could acquire the infrastructure and reorganize it into one system would control the modern Aegean's lifeblood — the cargo, the cars, the trucks, the people and their money — all moving ceaselessly between the seaports of the northeast Mediterranean. This ferry business was not glamorous, but it could be very profitable if you knew how to play the game.

The regulatory environment was the keystone, he summarized, including the domestic and international rules governing everything from the mandatory retirement age of ships to the number of life jackets stowed aboard each vessel. One might think that such regulations were not within the purview of those whom the rules would regulate, but this was not the case. Fortunately, Sfinias and Minoan Lines nurtured close ties to the political party in power, and together they worked toward regulations that made the plan all the more feasible. Perhaps the largest obstacle had been the new European Union rule that all open-ocean ferries be retired after 27 years of service. The ruling's intent was to get older, less-reliable and less-safe ships off the water and have them replaced with newer and faster models with stronger hulls, safer designs, and up-to-date lifesaving gear. But Sfinias had presented the Greek ferry industry's arguments articulately to their new European friends, pointing out the uniqueness of the Greek ferry business, its importance to the country's economy, its role in tourism and industry. Unlike operators in the English Channel or Scandinavia, Greek shippers had many more ports to service and a large population that depended on them for their very existence. Forcing Greek operators to retire ships after 27 years would bring total financial ruin to the industry, they argued. They could not possibly operate under such a rule, most certainly over the near term. The EU regulatory commission finally agreed and granted Greece a variance on the 27-year retirement-age regulation, deciding that — at least until the matter was reconsidered by the commission in another five years — Greek ferry companies did not have to scrap their ships until they reached 34 years of age. For Minoan Flying Dolphins and Pandelis Sfinias, the planning and positioning had paid off again.

With the EU regulators having taken the bait, Minoan Flying Dolphins was poised for the next step in the scheme: running available old ships on selected routes or, if they choose to do so, buying at bargain prices the old ferries now forced onto the market in other parts of Europe. A 27-year-old ferry, now worthless in the rest of Europe, could make money hand over fist for another seven years in Greece. One had to take advantage of every opportunity, especially if the opportunity was of your own making. This was business, after all.

In 1999, Minoan Flying Dolphins acquired 70 aging ferries and hydrofoils, getting the equipment at a steal. Yes, yes, they would modernize their fleet with new high-speed catamarans and super-fast ferries as they could be afforded, but high-volume and longer-distance routes would still be serviced with the older, larger, and enormously less-expensive ships. The pieces had continued to fall into place, and it had been due in no small part to the connections and untiring effort of Minoan Flying Dolphins' chief executive officer, whose emotionless façade at the meeting around the table in the plush office high above the street below concealed the anguish and utter desperation of the true man underneath.

❖❖❖

The summary of recent years continued. The last element of the plan required advances on the political front. "The Greek government," Sfinias had argued cleverly before, should "determine the rules of the game so that they are fair for all companies and EU countries, without making exceptions for Greek shipowners," casting an air of modern business objectivity to the discussion. But in his next sentence he made his real objectives crystal clear: "Yet, this doesn't mean that anyone can enter Greek waters on their terms. Rules will be implemented. Otherwise, who would maintain lines to the small remote Greek isles during the winter months?" "The Greek government... will establish rules, like licenses of expediency, labour contracts, etc., imposing certain obligations on shipowners, and in exchange the government will grant licenses." What he proposed was not a classic American approach in which the government defined the service to be provided and companies submitted their bids for the contract. Rather, Sfinias described an "arrangement," one in which there is "...an exchange between shipowner obligations and government control of price lists, routes and crew synthesis." It was, according to Sfinias, a "...system of issuing licenses under specific terms," a system that served the needs of the people but also the needs of the ship operators. In keeping with the longstanding and often-questioned ties between the shipowners and the government, individual shipping companies would service certain routes under "arrangements" with the government regulators. The result was that true competition, especially competition from foreign ferry companies, would be stalled — like the regulations for mandatory retirement of old ships — for five years, plenty of time for Minoan Flying Dolphins to establish a choke hold on the ferry market in the Aegean.

❖❖❖

With the background and history out of the way, he moved on to the current crisis and the events of the past two months. It had been a big fall from the heady days of one year before when their plan rolled on unimpeded like a massive swell sweeping across the open ocean. As everyone in the office knew, the *Express Samina,* a 345-foot 4,407-ton 34-year-old ferry, had sunk after 10:00 p.m. on Tuesday the 26th of September with 550 passengers aboard. Since then, a day had not passed without yet another shocking revelation from the television newscasters and the next public announcement from the government regulators. His life and that of others in the business had been threatened, bombs had been planted, lawsuits had been filed, employees had been jailed. It was true that perhaps he and others at the top had not paid enough attention to the day-to-day operations of the ferry business and the management of each ship and crew. But he had cultivated so many friendships over the years, and it pained him to see so many he knew turn against him and the company after a tragic accident like this. The latest rumors were that he — as the biggest fish — was next on the list for prison.

The newsclippings and preliminary reports on the table made it all painfully clear: the *Express Samina's* helmsman and midshipman were in the big room on the main deck behind the bridge watching the action-packed soccer match on the television with many of the 550 passengers as disaster was about to strike. It was one of the biggest games of the season, the match between the hometown team from Panathinaikos, near Athens, and the team from Hamburg. The white ball, bright green uniforms of the Greek players, and red shirts of the team from Germany made it easy to follow the game even if you were quite far from the screen. It had been an exciting match, and it wasn't easy to tear oneself away from the television when the home team was playing and there were so many feverish soccer fans on board. Captain Yiannakis had been watching some of the game himself, but now he was taking a snooze in his private room. He was supposed to have been awakened a dozen minutes before when they were many miles outside the harbor by first officer Psychoyos, who was technically in charge while the captain was napping but busy making the moves on an attractive woman in a far-off corner of the lounge. While the ball bounced around the field and the colorful players darted and dashed and the passengers and crew shouted their support for the teams and first officer Psychoyos devoted his attention to the pretty young lady, the *Express Samina* chugged on through the stormy sea at 19 knots, steered straight ahead by her trusty autopilot, all alone on the quiet and empty bridge of the ship. The setting was not lost on survivor Christa Liczbinsky from Germany, interviewed later by the reporters, who turned to her husband, a Lufthansa Airlines pilot, and remarked jokingly during the soccer game, "Who's driving the ship?"

Directly ahead lie the Portes islets, locally known as The Gates, the 20-meter-high twin rock spires two kilometers outside the destination port of Paros, marked clearly with fully functioning navigation lights. Suddenly, someone shouted out about an impending collision. The helmsman and first officer ran to the bridge and turned the wheel, but it was too late. Many of the 550 passengers watched out the windows in horror as they drove straight into the towering rocks. A side stabilizer blade well below the water line hit, ripping a 3-meter hole into the hull. The *Express Samina* came to a grinding halt and began to flood. The ship could survive a breach of the hull of this size, but the watertight bulkhead hatches had to be closed. However, the crew had not followed regulations and had not closed the hatches for the 5-hour trip. The water rushed throughout the bow and then back toward the stern. She began to list to port, the power was lost, and the lights went out.

The *Express Samina* began to settle slowly with her bow somewhat down but relatively level. No safety instructions had been provided by the crew when they left port five hours before, so the passengers were on their own. There was a scramble for life jackets. A crewman appeared and announced that everything was under control; the ship was in no danger of sinking, he said. There was no need to worry. Everyone should stay where they were, inside the ship, and await further instructions. But the one hundred or so foreign passengers who did not speak Greek did not understand the instructions, and most left the apparent safety of the large main rooms and went out on deck in the storm, prepared to abandon the sinking ship.

For 45 minutes the 4,407-ton ferry continued to settle, her open main deck drawing closer to the storm-fed waves each minute, the largest swells now drenching the terrified passengers out on the deck. One of the follow-on newsclippings had an interview with a tourist from Oxford, England. Like others, he had tried unsuccessfully to launch a lifeboat. Some of the equipment was missing and he could not understand the instructions which, although in English, were awkward and unclear. He had figured out some of it, but when he got to the part where he was to "cross two ropes" there weren't two ropes to cross. At this point the lifeboat was free from its mounting, but it could not be launched. He gave up and prepared to go overboard.

After 40 minutes the deck was down to the level of the churning water. More passengers jumped over the side, while others climbed over the rail and stood on the edge, waiting until the last possible moment, each wondering if they were jumping to survive or to their death. Most passengers on the open deck huddled about, not knowing what to do. The ship settled further down and the main deck of the ferry boat that was longer than a soccer field sank under. Perhaps a hundred people, most wearing old life jackets stuffed with cork and wood, bobbed to the surface as the waves washed them across the ship and the deck slipped away beneath them. Ironically, they were the fortunate ones and would all survive. As the upper decks of the ship submerged beneath the boiling surface and the *Express Samina* disappeared, the British tourist from Oxford found his lifeboat afloat, albeit upside down. He clung to it for his life, a modern-day Ishmael afloat on the carpenter's coffin.

The passengers, mostly Greek, who had followed the orders of the crew to stay seated indoors were unaware of the extent of the danger and had to escape quickly from the rooms as they flooded. People on the lower decks never had a chance. The deck-class passengers still in the main enclosed lobby had a bigger challenge according to the latest reports from the press: as was usually the case on the *Express Samina*, the big side doors on the main deck-class room may have been wired and chained shut to keep lower-class passengers out of upper-class areas.

❖❖❖

The press became insufferable in the days that followed, but, honestly, could he really blame them? Eighty-two people had drowned. Some of the bodies floated into the rocky coves of the island, videotaped by the news helicopters for all to see on the television reports. Others were found by rescue divers the next day, arms and legs of lifeless bodies waving gently in the undersea currents of the Mediterranean, life jackets hung up on equipment or stuck in the rails, ghostly open-eyed faces with gaping mouths trapped behind flooded panels of glass and locked doors. Quick-thinking fishermen and two British warships had picked up the survivors.

The captain, first mate, helmsman, and midshipman were arrested two days later on charges of "homicide with possible malice, causing serious bodily injuries with possible malice, violating maritime regulations, violating international regulations, ...sinking a ship, ...and deserting a ship." The captain had appeared in court on the island of Syros, handcuffed and humiliated, and was paraded before the cameras and the press like a man being led to the gallows before a bloodthirsty mob.

Minoan Lines had to respond to the negative press. After all, the financial health of the company might be at stake if people continued to get carried away with their emotions — and there was no shortage of emotion by September 28. They crafted and released a response to the press with the intent of clarifying the business relationship between the various arms of the company and minimizing overall financial damage, but it backfired. In retrospect it was not surprising that people took it the wrong way, seeing the announcement as a callous attempt to contain their monetary losses without mention or regard for the loss of life and responsibility. The press release read:

> "For your accurate and comprehensive information, Minoan Lines' management responding to the latest media reports following the sinking of the ferry boat *Express Samina* feels obliged to clarify the following: Minoan Lines is a minority shareholder of Minoan Flying Dolphins with a 31.6% stake; however, we wish to emphasize that there are no common business, managerial or organizational identities between the two separate legal entities. The ferry boat *Express Samina* belongs to Minoan Flying Dolphins, a company that shouldn't be mistaken with Minoan Lines, an autonomous legal entity and with a separate shareholder structure."

The fact that Pandelis Sfinias was the Vice President of the parent firm and Chief Executive of the subordinate company was conveniently omitted from the press release.

The aftershocks grew larger the first week in October. The government's preliminary investigations concluded that all Minoan Flying Dolphins' ferry boats were substandard. Greek Premier Costas Simitis addressed parliament on the accident, proclaiming that those responsible would be held accountable. The Communists and left-wing politicians cried out that it was the ruling party and their back room deals that had brought this all about. The government threw a counterpunch to quell the opposition, announcing that the Greek island ferry routes would be opened to full competition one year earlier than planned, a ruling that would benefit Minoan Flying Dolphins' chief potential rival, Strintzis Lines. Strintzis Lines, in turn, announced that it was raising 200 million dollars to be spent on new ships. Minoan Flying Dolphins' long-planned October 11 listing on the Athens Stock Exchange would have to be postponed, Sfinias and others had decided reluctantly.

The courts seized four of Minoan Flying Dolphins' high-speed ships on October 14 as guaranteed payment for unfolding legal suits. That same morning a bomb was planted outside the Piraeus office of the former minister of merchant marine affairs who was rumored to have been behind a number of shady ferry business deals. The anarchist group Revolutionary Nuclei claimed responsibility. Eighteen members of the *Express Samina* crew filed suit against the company on October 25, seeking 2.9 billion drachmas from Minoan Flying Dolphins, claiming they had been rendered "nervous wrecks" by the accident.

On October 31 the government approved a 10 percent increase in ferry boat fares, issued new ferry operating licenses to Minoan Flying Dolphins' competitors, and recalled the license of the *Express Naias*, the *Express Samina's* sister ship. November 8 saw another quarter-million dollar court award and the arrest of the Greek Harbor Corps Chief on charges of dereliction of duty with regard to inspections of the *Express Samina*.

The latest blow was particularly personal. Minoan Flying Dolphins had secured loans totaling 4 billion drachmas from EFG Eurobank and Alpha Bank one year before with Pandelis Sfinias' entire personal fortune as collateral. The deadline for payment was December 29, only a month away. With the company all but shut down and the onslaught of lawsuits, the cash had vaporized. They no longer had the funds to make the payment. Sfinias would have to pay up.

❖❖❖

He thought of his wife, Julie, at his estate on the island of Ekali, and of his sons. He thought of the 82 lives. These were his only regrets, his only silent pleas for forgiveness. Pandelis Sfinias, architect of a business empire, Chief Executive of the ferry company Minion Flying Dolphins, President of the Union of Coastal Ship Owners, multimillionaire Greek business executive well-connected to the most powerful politicians, pushed his chair back smoothly away from the table and stood up casually in his tailored suit, as if in thought about the next point of discussion at the summary meeting around the table in the plush office high above the street below.

He walked to the big window. It was now noon and the sun was even brighter over the blue of the Aegean Sea and the thousand islands of Greece. He casually pushed it open as if to take in the view and then stepped up and out and away without saying a word.

REFERENCES AND NOTES

A Greek shipping executive leapt to his death (2000). *Wall Street Journal*, November 30, 1.

Bomb planted outside ex-minister's office (2000). *Associated Press*, October 15.

Cadwalladr, C. (2001). Last year 82 tourists died when the Greek ferry Express Samina sank; what lessons have been learnt? *Daily Telegraph*, June 9.

Carrol, R. and Howard, M. (2000). Friends defend Greece's most hated man. *The Guardian*, September 30.

Carrol, R. and Howard, M. (2000). Slum boy who made good is now Greece's most hated man. *The Guardian*, October 1.

Court awards $250,00 (2000). *Associated Press*, November 11.

Court freezes dead executive's assets after shipwreck (2001). *Associated Press*, January 20.

Crew held in ferry tragedy (2000). *Santa Barbara News-Press*, September 28, A8.

Crew files lawsuit (2000). *Associated Press*, October 25.

'Dreadful boat' hits the rocks, killing 63 (2000). *Navigation News*, November/December.

Express Samina (2001). Ferry Management Services, Ltd., *FerryNews.com www site*, February.

Ferry captain admits he was napping right before accident (2000). *CNN.com www site*, September 29.

Former ND leader Evert files suit against former merchant marine minister, Minoan Lines, competition commission (2000). *Associated Press*, October 20.

Greek captain, crew testify (2000). *Associated Press*, October 3.

Greek ferry captain asleep (2000). *BBC.com www site*, September 30.

Greek ferry crew on murder charges (2000). *BBC.com www site*,

September 28.

Greek ferry described as 'a grime bucket' (2000). *CNN.com www site,* September 27.

Greek ferry executive suicide (2000). *Hellenic Times,* November 30.

Greek ferry rivals face up to different futures (2000). *Associated Press,* October 11.

Greek harbour corps chief charged (2000). *Associated Press,* November 10.

Hope, K. (2000). Greek shipping probe ordered. *inforMARE [online newspaper on transports, Italy],* September 29.

Howard, M. (2000). Head of ferry disaster company takes own life. *The Guardian,* November 30.

Interview with Pandelis Sfinias (1998). *Hermes Greece Today Magazine,* December.

Investigation continues in ferry sinking (2000). *Associated Press,* October 3.

Lowry, N. (2001). Lapses are blamed for Greek ferry. *Australian Marine Pilots Association www site,* October 3.

MDF accused of pressuring coastguard (2000). *Associated Press,* November 8.

No evidence of foul play in probe into Sfinias' death (2000).

Associated Press, December 6.

Papoutsis: the blackmail by the coastal shipping companies will not be accepted (2000). *Associated Press,* October 13.

Samina Captain: first mate was grossly negligent (2000). *Associated Press,* October 10.

"Samina" death toll rises; second ferry runs aground (2000). *About.com www site,* September 30.

Ship captain testifies to Nautical Accidents Council (2001). *Associated Press,* February 28.

Ten percent rise in sea fares (2000). *Associated Press,* November 1.

The disaster (2001). *GreekIslandHopping.com www site,* November.

The four high-speed ships of the Monoan Lines seized (2000). *Associated Press,* October 14.

The suicide of shipowner Pandelis Sfinias has shocked Greece (2000). *Associated Press,* November 29.

ABOUT THE AUTHOR

Steven M. Casey, Ph.D., is President and Principal Scientist at Ergonomic Systems Design, Inc., a human factors research and design firm in Santa Barbara, California. He received his formal education in psychology and engineering from the University of California and North Carolina State University. His work has covered a range of products, systems, and settings, including aircraft interiors, industrial control centers, nuclear power plants, oil-field machinery, portable electronic devices, automobile controls, agricultural and construction machinery, entertainment operations, and recreation and retail facilities. He travels extensively, and has completed projects for clients in the United States, Canada, the United Kingdom, France, Germany, Belgium, Denmark, Finland, and Japan. The objective of his work is to help make things, especially things involving advanced technology, easier and safer for people to use. He is the two-time recipient of the Alexander C. Williams, Jr. Award for outstanding human factors design contributions, and he has received a number of BusinessWeek/IDEA Gold awards for product design.